SPARTANAT BLACK BOOK

5

SPARTANAT

spartanat.com

Doktor-Herrmann-Gasse 4
9020 Klagenfurt am Wörthersee, Austria

ISBN 978-3-903526-17-4

DEFENSE AGAINST DRONES

DEFENDING AGAINST THREATS FROM THE AIR

COMBATTING SCOUT AND ATTACK DRONES

KRISTÓF NAGY & MARKUS REISNER

TABLE OF CONTENTS

A target in the sights of the "Lancet" — a Russian loitering ´-munitions—which has proven to be a great threat.

INTRODUCTION

Anyone who wants to know how dangerous drones are on the battlefield can go to YouTube or other social media channels. There, you will find plenty of videos from Ukraine made with and by drones, which document on the one hand what the drone weapon can now do, and on the other hand, show why it is so important to have a functioning drone defense. Because every weapon is never at home on just one side, and here too, the war in the East shows that there is a parallel evolution. Even if the Ukrainians are leading the way in using drones in combat, the Russians have learned quickly. Both sides must learn to deal with the constant danger.

The first drone video from the Ukraine war we saw was made shortly after the Russian attack. Young volunteers presented a small DJI drone and how they used it: namely as a scout at close range, sometimes even at ground level, to determine where the enemy forces were in the area. From that point on, the situation has escalated to such an extent that observers even speak of a "transparent battlefield" because drones are constantly on the move in surveillance mode and report every observation, which also makes it difficult for any attacker to achieve any kind of surprise effect.

Observation is already an essential tool. But drones quickly evolved into weapons, serving as bombers armed with hand grenades, rifle grenades, or RPG charges and dropping them. The videos are always astonishing, showing how rarely ground troops notice that an armed drone is hovering above them, which then drops its load with surprising accuracy into the circle of soldiers below.

Among the latest developments is the FPV drone, which was originally intended as a civilian "racing aircraft" which is controlled from the pilot's perspective using "first person view" glasses. The military drone operator flies a live explosive device directly into the target, which he can also track—from tanks to individual soldiers. The videos show the attack flight, which ends in the gray noise of the explosion on the screen. That sounds more "individual" than it is: excellent operators use up to 40 FPV drones per day. We have received confirmation that the Ukrainians have platoons of FPV drones, and there are reports they are also experimenting with an entire battalion of FPV drones and operators.

"But where the danger is, also grows the saving power," says the poet, and he is of course right. But developments in the drone segment are moving so quickly that countermeasures are lagging. People already saw civilian drones as a potential threat as missiles, even before they developed them into weapons systems. The first two means of defense that were seen were nets fired at close range and soon jamming to distract unwanted high-flyers with radiation, force them to land or cause them to crash. Laser weapons that can set a drone on fire in the air are still experimental dreams of the future.

The war in Ukraine has seen many instances of easily locatable DJI drones being used—typically with the risk of the operator being swiftly located and bombarded by grenade launchers or artillery within minutes. From shotguns to machine gun and anti-aircraft fire, we have seen everything in terms of drone defense that could be helpful in bringing the dangerous guided missiles out of the sky. The most visible is the installation of wire cages on vehicles and in front of gun emplacements to prevent drones from flying in. Drone defense is still a rapidly developing topic. The last, current chapter seems to be defense drones that act as "ram hunters" to bring down valuable enemies—larger observation drones in the depths of space.

With "drone defense" we want to give the user an initial orientation of what is possible. Christian Väth, an infantry officer, advises on tactical drone defense, Kristóf Nagy gives an introduction to the current spectrum of drone defense measures, Markus Reisner outlines the danger in "Kalashnikovs of the Skies", Gustav Freimann—himself deployed in the

Ukraine—reports on drone defense practices at the front. We hope that with this book we have made a start to the discussion of the topic of drone defense, which will inspire many readers.

April, 2024

TACTICAL DRONE DEFENSE

1

CHRISTIAN VÄTH

INTRODUCTION

An infantry squad hastily occupies its prepared positions in a combat trench. The soldiers are intently observing the battlefield in front of and above them. A drone operator has spotted movements near the enemy positions, which are several hundred meters away. Shortly afterwards, contact with the drone was lost. Perhaps an attack is imminent. Explosions caught their attention suddenly, but they were coming from far behind the squad, not in front of them. Columns of smoke rise where the battalion command post had recently been. Then silence again. The Squad Leader is the first to hear a quiet whirring sound. Then the machine gun squad hears it, too. It gets louder. Panic rises in the soldiers as their bodies release large amounts of adrenaline. Some get goose bumps, others have cold sweats running down their necks. The squad, with their battle-hardened experience, has inflicted losses on the enemy and endured their own. One of the younger men, however, spontaneously empties his bowels. Everyone here knows what is coming next.

They still can't see anything. The noise grows louder, and now it's so close that the squad leader can tell it's actually many individual sounds merging into a high-frequency hum. About a dozen. On the left flank, the positions are in a small patch of forest, and visibility is short. Further to the right, it's more open. From there, the squad leader looks up into the sky above the trees. Through the tops of the bare branches, one can see small black objects moving very quickly. He hears himself yelling "DRONES!!!" at the top of his lungs, but it feels like he's standing next to himself, watching himself. Everything is automatic. His hands are shaking. Two riflemen fire over the trees, but they miss. It's more of an act of desperation, giving them the feeling that they can do something, somehow regain control. The drones are now clearly visible, racing toward the trench. Finally, the shotgunner is in a suitable position and opens fire. But he only hits one target. There are simply too many. Everyone makes themselves as small as possible, seeks cover to avoid the inevitable impacts of the FPV drones. It's personal. These drones are on the hunt and usually hit their victims like bullets. But not this time. The dozen fly over the trench at high speed at a height of just two to three meters. Are they not the target after all?

The Squad Leader looks in horror at the barely audible black dots. Then an explosion knocks him off his feet. The fighting position a few meters away from him has been hit by a larger caliber projectile. Further detonations occur all around, including in the neighboring squads. Anyone who is too close to the impacts hears nothing but whistling. The Squad Leader has the machine gun watch the area in front, while some riflemen tend to the wounded. There are at least two of them. He looks into the trench that leads to the platoon command post. There he sees the platoon leader and his radio operator, both running towards him. The platoon leader is obviously shouting something, but the Squad Leader only sees the movements of his mouth, he cannot hear him—the whistling in his ears is still too loud. Then he sees the drones in the background. They are coming back! He tries to give the platoon leader a hand signal, but then they are already there. He can no longer resist the urge and throws himself into the nearest cover. Approximately half of the drones targeted various spots within the squad's position. The platoon leader dies instantly.

The machine gun position was badly hit. The Squad Leader tries to restore order and man the machine gun. Several large explosions occur on the right flank. In this chaos, no one notices the small reconnaissance drones that are hovering just 100 meters away and providing the artillery with target data. A rifleman yells “TANKS!!!”. Two armored personnel carriers are visible at full speed on the right flank. Perhaps the armored infantry have already dismounted and are fighting in the positions of the neighboring squad. One tank bursts into flames; the Squad Leader does not know what hit it. Only three men from his squad can still fight. He searches the churned up earth for the anti-tank weapons. A hand grenade suddenly lands next to him. He sees exactly how it falls into the loosened ground. Then everything goes black. A drone has dropped the grenade directly above him. The Squad Leader is dead. Despite their dogged fight, the last defenders are ultimately overrun, and the enemy breaks through the first line of defense.

BASICS

Our combat example is fictitious but could have happened in every detail. The establishment of nano and small drones as universal weapons of war for reconnaissance and impact on the modern battlefield has created an enormous need for defense capabilities everywhere. What is new here is the required density at the lowest tactical level, which is becoming a necessity because of the mass use of inexpensive drone technology. Just as with unmanned systems, this has also led to a sudden increase in the variety of brands and production capacities on the global market in drone defense within a brief time.

The areas of application are broad and range from infantry squads in combat to protection against sabotage of critical infrastructure at home. The range of technical solutions from smaller portable devices to strategic air defense is now very large. Therefore, in this work, we want to concentrate on the defense against nano and small drones by individual users or small troops. Those requiring such equipment are essentially dismounted combat troops, special forces, internal security organizations and, in the future, possibly also civilians.

In the following, we look at three standard scenarios that include all the major challenges of light and mobile drone defense. The limiting factors are always weight and dimensions, as all users already have to carry many items of equipment and supplies to fulfill their mission. Robust construction, ease of use and excellent suitability for everyday use must, therefore, be the key features of corresponding solutions. Because of the established use of drone technology in conflicts all over the world, it will also have to prove itself in real-world use in the short term:

- **Has the system already achieved real defensive successes?**
- **If so, to what extent and under what conditions?**

DETECTION

Let's turn back the clock of our nameless Squad Leader a few months. He and his men are moving in a line through a densely overgrown forest area. Visibility is about 30 meters. It is summer and the canopy of leaves protects them from being detected from the air. Nevertheless, the squad is moving slowly and carefully. A vehicle dropped off the squad about two kilometers earlier, and now they have to continue on foot to reach their own positions. There they are to relieve another squad that urgently needs to be refreshed. There are two critical points on their route: a path that has to be crossed halfway and the last few meters to the trenches. At these points, they will be vulnerable and have no visual protection.

The Squad Leader lets the rifleman go first, followed shortly by the machine gun team and himself, then the rest. In this situation, reconnaissance drones are to be expected in particular, which provide the artillery with target data. If the lane is actually being monitored by a drone, the pilot must keep it relatively low in order to see details at the edge of the forest. At a low altitude and with little movement, the shotgunner has a chance of hitting the drone. As soon as the lane comes into view, the Squad Leader has everyone stop for a moment. Everyone tries to see or hear something. Nothing. Every few minutes, we hear impacts, but these are far away on a

We need compact systems for protection while moving.

different section of the front. A few craters already marked the surrounding forest. Thankfully, there are no more mines here. "It would be better if we could go further to the right and avoid the lane," thinks the Squad Leader. But you rarely get to choose your paths and borders on the battlefield. They have to get through here now.

There are two ways to get to the other side undetected: slowly, with the best camouflage, or in a quick, closed-ranks rush. There is no time for the first option, and the Squad Leader prefers to test the men's patience elsewhere. He has to consider when and where he will demand maximum discipline from them. If they jump quickly now, they will also have to run the rest of the way. Because if they are seen doing so, they will have very little time before the first artillery shells can hit. After the lane, there are still about 400 meters. He gives the necessary hand signals. The squad must keep its signature as low as possible: no unnecessary movements, no reflections, no noises, no conspicuous traces. Now everyone is ready to jump. The Squad Leader counts down silently with the fingers of his hand: 3, 2, 1.

Everyone runs as fast as they can. They have learned to concentrate on their target and to get there as quickly as possible. But almost all of them find themselves looking worriedly towards the sky. As if it would change anything to see a drone that might be hovering up there. Nobody sees it. But it is there. A drone, the kind you can buy anywhere, is hovering about 70 meters above the treetops. It sends its images in real time to the screen of the enemy pilot, who is sitting behind the enemy trenches. An experienced soldier who has been flying in this sector of the front for weeks is at the remote control. Another drone had spotted the vehicles that were transporting our squad. It was then clear to him that either supplies or relief were coming. Either way, there would be targets. He achieved his reconnaissance aim by performing a “fishhook” maneuver. To do this, he flies at maximum altitude in a wide arc around the squad in the front trench in order to get behind them. Exactly where the relief has to go. Now he sees them rushing from behind, and they don't see or hear his drone. He has already prepared the target data and passed it on to the artillery. Now he just triggers the fire: Go!

The Squad Leader tries to keep an eye on everyone while they hurry. On the other side, the forest is much sparser, mainly because of previous impacts. They keep running. Hopefully, no one stumbles. Hopefully, the leader of the group in the trench has shared the news of the relief with all the team members. They will know in a moment. The vegetation becomes increasingly sparser until the "trees" are nothing more than bare sticks sticking out of the ground. They are now running across completely open terrain. Why do we, of all people, have to go into the trench that is the hardest to reach on the entire front? The men's lungs burn. They'll almost make it. There! Soldiers from the other squad signal at three different spots in the trench, indicating that everyone should move to the nearest spot to them. The Squad Leader can already see the face of a soldier, he is happy about the relief. Only a few more meters. At the trench, he does not jump in, but turns around. He wants to see whether his entire squad has stayed with him. He lets the machine gun squad into the trench first. It looks as if everyone is there. Just as he is about to turn around, he sees an enormous fireball and is knocked off his feet. He falls backwards into the position. More impacts. Someone is pulling on him, but he just tries to make himself as small as he can. He is grabbed by his arms and legs, and now he looks up briefly. His comrades want to pull him into a fortified shelter. He crawls forward into the dark hole in a flash. There is dirt and dust everywhere.

The basic problem on the transparent battlefield is the signature. Every soldier, vehicle, and drone creates a multi-layered signature, which can be detected in several ways. And a unit on the ground can usually only detect small unmanned systems if they focus all their capacities on them. However, if they fully focus on looking for drones, they cannot carry out their actual mission. Therefore, tactical movements have become so difficult even for small units, because they can't be fully aware of drone threats if they are also on the move. Therefore, combat troops need a protective umbrella above them that can detect and ward off threats.

Certain timeless tactical truths also still apply in the era of drone warfare:

THE BASIC PROBLEM ON THE BATTLEFIELD IS THE SIGNATURE. EVERY SOLDIER AND EVERY DRONE CAN BE DETECTED.

▸ INDIVIDUAL DEFENSE

The density of sensors on the battlefield makes areas that offer a lot of privacy more valuable. This is almost always the terrain in which light infantry forces have their highest combat value: forest areas and urban areas. The many obstacles and short visibility distances in such terrain negate the classic strengths of armored combat troops. Such terrain also limits the operational value of drones.

Our squad has fended off enemy attacks in various defensive positions over the last few weeks. The entire battalion has now been relieved from the front and has dedicated two weeks to training replacement personnel and receiving new material. Now our squad finds itself at the center of the fighting, in the middle of an attack operation, to take a tactically important small town. A shock troop infiltrated the enemy defenses at night and create a breakthrough point through which as many forces as possible must now advance as quickly as possible. The enemy knows this too and is concentrating its drone forces and artillery fire on this spot on the outskirts of the town.

The Squad Leader had to split his men between two vehicles; there is not enough room for his twelve men in an armored personnel carrier. He has to bring a drone squad to the commander on site, which has resulted in two more men than usual being present.

This will be a welcome reinforcement for the assault squad at the front. It is loud and hot in the armored personnel carrier. Because of the commander's fear of being hit by an FPV drone, the vehicle is driving fast. The men hear the signal over the on-board loudspeaker. They will soon be dropped off directly at the point of entry. The tank suddenly stops, the rear hatch opens. Bright light penetrates the interior. The Squad Leader does not have to urge his soldiers on, everyone wants to get out, to feel solid ground under their feet. They carry a heavy load, with almost everyone gripping ammunition boxes or hand grenade containers in both hands. The order was to take as much as they could carry. The crew of the armored personnel carrier throws more boxes after them before the rear hatch closes again.

The Squad Leader has to get his bearings for a moment, but quickly recognizes comrades in a ruined house, frantically waving with a safety vest in their hand. Without a word, he simply runs towards the ruins. Everyone else follows. Three men come towards him and run past him—they are getting the ammunition. Before they even reach the pile of rubble, which was probably once a single-family house, the armored personnel carriers are already on their way back. Suddenly someone shouts: "DRONE!!" The Squad Leader only hears the whirring of an FPV drone for a very brief moment, then there is an explosion on the other side of the building. Two men get hit, one is seriously injured. It doesn't look good.

While the squad supports the wounded, the Squad Leader gets a brief by a comrade about where their own forces are in the destroyed town and where the enemy is. The squad takes the wounded to an aid post in the nearest basement. The squad then sprints past a few houses to a former supermarket, where they report to the platoon leader. Shortly before reaching their destination, grenades, perhaps mortars, start hitting the surrounding streets. The squad seeks cover in the next building. After the last impact, they wait briefly to see if there are more to follow. At that moment, a rifleman notices a drone above a gas station on the other side of the street. It is not moving; it is probably watching the platoon in the supermarket! The riflemen immediately open fire, but the target is too small for them to hit. The shotgunner carefully takes aim at the front door frame and brings the drone down from the sky with his second shot. Now they have to be quick. They have to get to their comrades in the supermarket immediately; another barrage may follow in just a few seconds.

Classic repeating shotguns are widespread on the Ukrainian front and have become part of the armory of most infantry squads. Although classic repeating shotguns are excellent tools, infantry squads will eventually require more powerful systems. A dedicated drone gun is not suitable for combat troops because it requires another user to carry it. In the medium term, technical progress will make small, compact attachments the size of a laser light module for the rifle possible, which will enable the shooter to fight drones. Until then, compact fire control sights such as the SMASH can ensure initial capability. Above all, however, combating aerial targets must become part of broad shooting training.

Almost like pigeon shooting: buckshot can be a solution.

▸ SWARM DEFENSE

Let's now rethink our combat example from the beginning of the chapter. This time, the squad has the support of a passive protective shield that can quickly disrupt the use of drones. In addition, the troops have an integrated mounted and dismounted drone defense.

The soldiers are intently observing the battlefield in front of and above them. A drone operator has spotted movements near the enemy positions, which are several hundred meters away. Shortly afterwards, contact with the drone was lost. Perhaps an attack is imminent. Suddenly, more explosions are heard, not in front of the squad, but far behind. Columns of smoke rise where the battalion command post had recently been. Then silence. The Squad Leader is the first to hear a quiet whirring sound. Then the machine gun crew hears it as well. It gets louder. The machine-gun is mounted on a compact carriage, with remote control, while the crew is sitting safely in a trench. The machine gun has a fire control sight that enables it to engage fast-moving targets. Some get goosebumps, others have cold sweat running down their necks. The crew, having experience in combat, has already inflicted casualties on the enemy and endured casualties themselves.

But the troops still can't see anything. The whirring noise grows louder. Then they hear a few quick, dull thuds, and they see small flashes in the sky above the treetops. The trench erupted with loud cheers. In quick succession, the entire sequence repeats itself three times. The platoon's machine gun has destroyed eight FPV drones on approach before now the gun has to be reloaded. Four of the FPV drones are still moving quickly towards the squad's positions. The machine gun crew has already taken aim and opens fire. Two more flashes. The two drones that get through race across the trench but find no targets. The troops all are in protective covered positions and observe the area via cameras or periscopes. One drone races on, but the other hits the machine gun. The squad's main fire support weapon is now useless, but at least there are no casualties. Another burst of fire from another machine gun, more flashes in the sky. Then, after a brief silence, the Squad Leader orders everyone back to their positions.

Chains rattle. Things are looking worse on the right flank: several drone swarms have overwhelmed the defense, enemy armored personnel carriers are on the attack to exploit the success. The Squad Leader forms two tank destruction squads, one with loitering munitionloitering munition and one with anti-tank grenades. The drone pilot supports the enemy with enemy reconnaissance. A short time later, four burning armored personnel carriers are on the battlefield. The enemy breaks off its attack. The neighboring squad suffered losses, but the front has held firm.

CONCLUSIONS

Because of the current limitations of the technical capabilities of portable drone defense systems, we must consider this form as the last resort of a multi-layered defense. Small dismounted units have a much better chance of successfully fulfilling their mission if they operate under a protective umbrella according to the classic principles of air defense. This includes rapid and as complete as possible detection of everything from individual nano-drones to large swarms.

Besides reconnaissance, real-time communication between the air defense and the front line must be possible in order to ensure a successful defense at all. This is hardly conceivable without AI support, especially with swarm threats. Early reconnaissance and effective protection of the airspace through electronic countermeasures leads to a decreased reliance on kinetic means in the final meters.

The focus of the defense organization must always be on the goal of avoiding a selective overtaxing of the troops on site. If the enemy breaks through with a massive use of these inexpensive systems, he creates the conditions for a breakthrough on the ground. An integral part of any drone defense should therefore not "only" combat enemy drones and loitering munitions, but also appropriate own weapons in reserve that can plug such gaps.

CLOSE-RANGE DEFENSE—SHOTGUN VS. DRONE

Simple repeating shotguns were among the earliest defense systems in the fight against drones at very short distances. But can you really reliably fend off drones with a shotgun? Or is it more of a myth? Experience from the first two years of the war in Ukraine shows that, at least in some sectors of the front, infantry squads have repeating shotguns across the board. But which shotguns are best suited, how can we tactically integrate them, and what ammunition do we need?

GEN. 2 SHOTGUNS ARE IDEAL

We can divide shotgun systems into four generations: simple single-shot guns like over-and-under shotguns (Gen. 1), repeating shotguns with tubular magazines (Gen. 2), self-loading shotguns with tubular magazines (Gen. 3) and self-loading shotguns with box magazines (Gen. 4). Each type has certain advantages and disadvantages.

A combat shotgun with the additional task of drone defense should be a Generation 2 weapon. Classic pump-action repeaters of this type are, for example, the Mossberg 590 or the Remington 870 series. With up to nine cartridges, they have a significantly higher capacity than single-shot guns, which even self-loading Generation 3 and 4 cannot surpass. When different cartridge types are required, this generation surpasses a self-loader in suitability and reliability by enabling quick changes of ammunition. In addition, experience has shown that different case lengths lead to malfunctions even in high-quality self-loading shotguns. Repeating shotguns are extremely trouble-free—as long as there is no operator error. However, military use of shotguns is not very common in Europe and usually only for special tasks such as breaching.

THE RIGHT AMMUNITION

Contrary to popular lay opinion, the Hague Conventions on Land Warfare or the Geneva Conventions does not ban the use of shotgun ammunition against persons. Two types of ammunition particularly interest infantry shotgun shooters: buckshot (00 buckshot) and shotgun slugs. A tried and tested repeating shotgun combined with high-quality buckshot enables a spread the size of a DIN A4 sheet of paper over a distance of 25 meters. Depending on the situation, it is possible to use it at longer distances. Since buckshot makes the most of the shotgun's advantages, shotgun shooters should always carry a weapon loaded with this ammunition and only switch to other types when necessary. Learning to change ammunition is easy for shotgun shooters and they can practice it easily in dry training.

In military scenarios involving regular infantry, the consideration of civilian collateral damage is far less of an issue than in special forces or police operations. Since we can accept greater dispersion here, fire is possible and effective up to around 50 meters. Despite the unlikelihood of buckshot penetrating an opponent's body armor, it still releas-

es an enormous amount of energy. The dispersion means that there is a very high probability of a first hit and the possibility of hits outside the protective plates. To give the shotgunner more options in more open terrain, shooters can use slugs. In combination with good sights, hits up to 150 meters are possible.

Birdshot is effective against drones, and the author has personally shot down a drone using a shotgun with such ammunition. Birdshot offers a higher chance of hitting the target, and compared to its impact on human opponents, it takes very little kinetic energy to disable a drone. So far, militaries have avoided using special anti-drone cartridges, as they are only suitable for very limited civilian use. In principle, one can only carry out defense at shorter distances (around 80 meters), and it will only be possible to land lucky hits—if at all—against fast FPV drones. However, embedded in a multi-layered drone defense of all troops, the shotgun can still be an important item in the toolbox.

SHOTGUN SHOOTER TACTICS

In military circles, the shotgun is completely misunderstood, especially in German-speaking countries, where they have used short assault rifles or submachine guns for comparable tasks. However, in the hands of an infantryman, the repeating shotgun can be the most effective firearm at close range. It always shows its advantages where the distances are short. Unlike pistols and rifles, the shotgun fires up to nine well-grouped projectiles toward the enemy with each shot. In contrast to the previous practice of carrying a special short shotgun just for breaching, the assignment of a dedicated shotgunner at squad level is not only promising but has already proven itself in numerous operations.

Whether in dense forest, in a trench, or in close-range urban combat, the shotgunner ideally suits the role of the first man (point man). Quickly suppressing an enemy that appears in the bush or in a window is the flagship role for this weapon system. If the infantry squad has to fight in open terrain—where it should not naturally be—the shotgunner can take on one of the many other tasks that always arise in combat. It is also rare that all

squad members will be involved in all firefights. Since the shotgunner does not have to carry magazines, but reloads loose ammunition, he could also carry a compact grenade launcher, for example. This means he can also fight effectively in open terrain and provide fire support to the riflemen of the squad.

In addition, the shotgun is a firearm that is widespread throughout the world and ammunition is available everywhere. It is therefore quite easy to procure one quickly for combat—even buying one privately if necessary. This aspect should not be neglected: both in Ukraine and in the fighting in the Gaza Strip, soldiers use privately procured firearms on a large scale.

KEY TASKS IN THE DEFENSE AGAINST SUAS

1. LOCATE

Find the sUAS in the airspace

2. LOCK

Determine the sUAS coordinates

3. IDENTIFY

Is it a friendly or an enemy sUAS?

4. TRACK

Follow the sUAS's movements

5. DEFEAT

Undertake countermeasures

COMBAT AGAINST SMALL DRONES

2

KRISTÓF NAGY

THE NEW BATTLE FOR AIR SUPREMACY HAS BEGUN

For decades, Western armed forces have become accustomed to conducting operations with complete air superiority. If this was not the case in the first few hours or days, as in the war against Iraq, then Western forces rapidly established complete air superiority by the start of the ground operation. However, the massive spread and availability of comparatively inexpensive small drones—known in military circles as **Small Unmanned Aerial Systems (sUAS)**—has added a new facet. In the 21st century, this leaves the use of the third dimension open to actors who, until recently, would have failed for both technological and cost reasons.

This (partial) loss of air superiority affects not only the consideration of near-peer conflicts. Even non-state actors can now pose major challenges for modern armed forces, especially in the short and very short range. The conclusion is sobering: **Western armed forces have already lost air superiority in the near-ground area before the start of the next military conflict.** This painful realization must be at the beginning of any consideration of the topic of "defense against small drones". A long overdue rethink is also taking place in the area of sUAS use. In the future battlefield, many unmanned aircraft will fill the airspace and collaborate with each other and with various manned and unmanned ground elements to accomplish their respective missions. In fact, U.S. Army and Marine Corps doctrines now state that first-contact in the future battle will be drone-on-drone rather than human-on-human.

Reconnaissance, target identification and tracking, deconflicting, and defense must take place in this area of tension. This is a tremendous challenge, which is also subject to rapid change with ever faster technology cycles. The fact is, **on the battlefield of the future, every soldier,** regardless of his or her function, **will confront enemy micro-drones (sUAS) at some point.** Therefore, we must think of defense against such systems holistically and not just limit it to individual units or means of action. A comprehensive awareness of the dangers and best practices must be created in order to assess and mitigate the threats posed by these systems.

An essay on the subject recently published by the U.S. Naval Institute bears the programmatic title "Make Every Marine a Drone Killer". This call is based on the realization that sensors and effectors are only tools and that drone defense for all troops is ultimately a question of doctrine and training.

WHAT IS SUITABLE, WHAT IS NOT, AND WHY?

The use and adaptation of preferably light, **short-range anti-aircraft weapons**, known internationally as "Short Range Air Defense" (SHORAD), some of which have been in use for decades and have proven themselves, against micro-drones seems like an understandable reflex. However, this approach is not suitable for many drones classified as "Class I" for several reasons. In the example of guided missiles, the cost of the defensive weapons compared to the target is disproportionate. This was demonstrated in Ukraine by the use of MANPADS against reconnaissance drones such as the Russian "Orlan-10". It has also been observed that drones are sometimes intentionally deployed near anti-aircraft systems to incite them to open fire and gradually deplete the system's combat load. With micro-drones, the cost disproportion mentioned above is even more extreme. The signatures of sUAS are also so low that even detection cannot occur early enough to deploy anti-aircraft missiles. The lack of an infrared signature or the radar cross-section that is too small also means that targeting is not possible. Modern drone defense systems that use missiles as a hard-kill option therefore often rely on laser target illumination, such as the U.S. "Vehicle-Agnostic Modular Palletized ISR Rocket Equipment" (VAMPIRE). For cost reasons, the effector comprises an unguided Hydra-70 rocket equipped with a control unit. Using such an effective agent against a tactical UAS, such as the "Orlan" family mentioned above or even Russian "Forpost" systems, i.e. drones of the "NATO Class II", seems to make sense. Systems such as the VAMPIRE are also not suitable and cost-effective means of combating multi-copter drones, which, even with their weapons load, cost less than 20,000 U.S. dollars.

The above considerations lead to the conclusion that **anti-aircraft artillery (AFA)** could be an effective means of defending against small drones. Depending on the caliber, the unit price per cartridge fired is reasonable at first glance, even if a higher ammunition rate is required for a reliable and devastating hit. In addition, the systems can be employed against both unmanned and manned aircraft, depending on their ballistic parameters and caliber. The successful use of "Gepard" anti-aircraft tanks against Iranian "Shahed-136" drones in Ukraine demonstrated this. In addition, Ukraine has used optronics to upgrade machine guns and cannons, some of which are decades old, to serve as anti-aircraft systems in the caliber range of 12.7 to 23mm. With such systems, they have also achieved a considerable kill rate against the "Shahed" (which is used more like a flying bomb than a drone by the

Russians). The fact that the "Shahed-136" was originally developed as a target drone to train anti-aircraft batteries adds a certain pleasant irony to the situation.

Besides implementing day and night combat or all-weather capability into the anti-aircraft systems, their inclusion into an efficient early warning system and the parallel use of electronic warfare means also plays a key role in the concept's success. Interestingly, the Russian armed forces seem to have partially adopted the Ukrainian concept and broken it down to the tactical level. Besides ZSU-23-2 machine guns mounted on trucks, electronic warfare equipment and smoke generators are also part of the Ukrainian light and mobile aerial drone defense forces. In particular, multispectral line-of-sight interruption using dynamic smoke screens can be an effective means of decimating the combat value of enemy drones. Therefore, all troops at lower tactical levels should include portable and hand-deployable smoke grenade systems in their drone defense armory.

It is also not surprising that the U.S. Marine Corps is testing weapons with calibers ranging from 5.56 and 7.62 up to 30mm for the "Marine Air Defense Integrated System" (MADIS) drone defense system. However, the supposed cost-effectiveness of AA systems is deceptive. The price of a representative MADIS system is comparatively high, and the number of deployed systems cannot ensure coverage in depth. Attempts to rely on vehicles equipped with light or heavy machine-guns for drone defense can only ease part of this deficiency, but not eliminate it. In addition, AA systems depend on the mobility of vehicles. Terrain contours, buildings or vegetation severely restrict these, as well as the target and effective range.

High-tech solutions with laser or microwave effectors, such as those currently being tested or developed, will not be available in sufficient numbers in the foreseeable future to provide a broad protective umbrella in the event of conflict either. If it is accessible, the **jammer option,** currently used as a solution for difficult-to-hit sUAS, can be employed. And although it is a workable option for protecting base camps or pre-planned patrols, an interlocked battle in urban or wooded terrain would significantly reduce its effectiveness.

In order to close the gap that has clearly already arisen, effective solutions are required as soon as possible at the lowest tactical level. We also need to consider force protection as a challenge that all branches of the armed forces must address. Besides an infantry squad, a ship or an airfield, as well as a fuel depot, are potential targets for small UAS. However, reality already shows that the range of targets that would have to be protected against reconnaissance, fire control and direct attack by micro-drones has become even greater.

The war in Ukraine has impressively shown **the use of unmanned systems down to the level of the individual gunner.** One possibility is the jamming technology already discussed. However, this soft-kill option is not available across the board and will not be in the foreseeable future for cost reasons. The operation of the systems and their size pose insurmountable challenges for small, but sometimes self-sufficient units in terms of transportation capacity and, above all, qualified personnel. Automation and miniaturization could remedy this in the future and provide a portable solution that ensures the required coverage. Even with these as yet intangible solutions, jamming as a means of drone defense is always associated with a clear electromagnetic signature from the effector. To avoid this on the battlefield of tomorrow, a passive solution is necessary for location and fire control. The former is possible, for example, through purely passive radio frequency detection. Unlike an emitting radar device, this mode of operation does not generate any radiation. The operation also involves lower power consumption, which is not insignificant given the energy hunger of modern system networks.

DEFENDING AGAINST AND DEFEATING MICRO-DRONES: TAKING THE FIRST BABY-STEPS

The primary focus of early drone defense systems was simply detection and tracking. In addition to their limited functionality, these systems were primarily intended for police use and could only process a few contacts. Such systems quickly reached the limit of their capabilities when they had to deal with even small swarms of just a few drones.

As drone technology progressed, people soon recognized the importance of correctly identifying and categorizing the detected contacts, and they needed to implement upgraded defensive solutions across the board. This is not surprising, because, as in civil aviation, for example, the identification of flying objects still relies on the transponder data actively sent by the aircraft. But the increased electronic signature an aircraft would produce prohibits the transmission of IFF transponder data on the battlefield. The problem remains distinguishing between your own drones and enemy drones. However, trying to establish IFF purely based on flight paths and vectors has a huge potential for error and is becoming increasingly difficult as the airspace becomes more saturated. Besides the potential danger of shooting down one's own sUAS, accidentally opening fire or using electronic jamming equipment would also reveal the position of one's own defenses and make them more vulnerable to enemy counter-fire.

Deconflicting between our own and enemy drone forces, as well as **airspace management,** are therefore future areas of responsibility that involve both the use of friendly unmanned aircraft and the defense against enemy drones, and require close cooperation between all air and ground elements. Developers have been working on automated systems for some time now that rely on artificial intelligence and passive detection to function as airspace scouts. Examples of this are the projects "Advanced Recognition Tool using Electromagnetic waves for Identifying unmanned aerial Systems" (ARTEMIS) and "Drone Identification System" (DroIDs) funded by the NATO Communications and Information Agency. The stated aim of both is to provide an inexpensive approach by using commercially available components to restore the cost balance between target and countermeasure. Both projects also aim to identify and categorize suspect unmanned aircraft using machine learning in order to increase speed and accuracy of decision making and to reduce the cognitive and logistical burden on troops.

Systems such as these two are being tested and validated by NATO every year as part of the "Counter-Unmanned Aircraft Systems Technical Interoperability Exercises" (C-UAS TIE). The transatlantic alliance recognizes the threat posed by sUAS, and the importance of sharing the data gathered from these exercises within its own organizations and with industry partners. A common effective solution is nonetheless still a long way off.

Detection systems currently available on the market also focus on detecting and identifying the communication signals sent by the drone or the operator. Drones that operate autonomously or on a self-guided basis are a tougher challenge. In the latter case, there are no control signals directed to the UAS. Instead, the drone flies along pre-programmed waypoints using GPS passive signals. However, their combat value is limited because they cannot transmit real-time data to a ground station, and some modern detection systems can even track down small drones that have autopilot control. Systems in this category are already available that are compact and consume moderate amounts of power. They can also carry out their task fully automatically and fit easily onto a Soldier's vest or plate carrier.

THE MINDSET FOR DEFENSE AGAINST DRONES

From the mid-1940s until well into the 21st century, armed forces around the world adapted step by step to the increasingly serious threat posed by low-flying **fighter-bombers.** In the 1970s, this threat was increased by the arrival of armed **combat helicopters.** In this environment, improving detection and tracking, raising awareness about the increased risk, as well as enhancing defensive effectors all played an important role right from the start. We need to replicate this process of learning and adapting in relation to the defense against small drones. The U.S. Army's **"ATP 3-01.81 Counter-Unmanned Aircraft System (C-UAS)"** regulation issued in 2023 demonstrates what this holistic approach can look like. The freely accessible doctrine examines the topic of defending against small unmanned aircraft systems at the brigade level and below. With this document, the U.S. Army responded to the changed threat situation, which makes long-lasting, high-intensity conflicts in the alliance environment more likely and for which the armed forces must urgently prepare. It is therefore advisable not only for military specialists, but especially for people interested in security policy, to deal with this complex of topics.

The "ATP 3-01.81" comprises four chapters and two supplementary appendices. The first chapter explains what unmanned aviation systems are in this context. Besides explaining different types and classes of drones, the first chapter also covers navigation and target guidance, as well as telemetry guidance transmissions. In addition, the chapter covers loitering munitionloitering munition systems as a separate means of action. The second chapter looks at planning aspects, primarily around points such as airspace surveillance and reconnaissance. Chapter three follows immediately after this and highlights the passive measures in the event of a detected UAS threat to one's own unit. The steps explained here, such as camouflage (visual and electromagnetic), dispersing and constructing fighting and command positions, are virtually identical to current measures for reducing the effect of an aerial attack by manned aircraft. The operational reality from the Ukraine war, however, shows that there are additional requirements for building and camouflaging fighting positions on the modern battlefield. Many of these tips and techniques are familiar from the catalog of measures taken to protect against snipers. For example, trenches and windows covered with nets can disrupt reconnaissance drones and also prevent attack drones from penetrating. As in defending against snipers, it is crucial to implement camouflage and blocking measures that span a multi-spectral range, not just the visual range seen by the naked human eye.

The U.S. Army document also highlights the activities of the airspace scouts and procedures for raising the alert, as well as techniques for combating drones at close and very close range with small arms and jammers. The fourth chapter goes deeper into methods of combating small unmanned aircraft systems. Besides defending against drones in flight, the focus is also on detecting and combating their launch facilities, ground stations and operators. Two appendices conclude the publication with one focused on planning and training for defending against drones, and the other being a presentation of currently used defense systems.

The aim of the ATP 3-01.81 regulation is to identify the threat to the country's own operations from unmanned aircraft and to show possible ways of defending against them. It is explicitly understood that combating unmanned aviation systems is not an organic task of certain weapon systems or branches of the military. Instead, every soldier and every unit must constantly carry out defense against drones and loitering munitions as part of the standard operating procedure and security tasks in their own area of responsibility. Each level contributes to a multi-layered defense by providing an additional line of defense that contributes holistically to increasing the survivability of all own and allied forces. This multi-layered defense should ideally comprise a combination of active and passive measures that complement each other.

Besides implementing immediate defensive measures, we also prioritize risk assessment and prior reconnaissance to create an ideally comprehensive situation picture. When developing such measures, one must also understand the perspective of the drones against which they want to protect themselves. Previous standards and practices for camouflage protection against aerial observation, for example, are not as effective against agile small drones. The spectacular images of FPV drones from the Ukraine war, which fly into buildings and specifically attack vehicles or individual soldiers fighting in positions, show this. A change is required here. Military leaders use the tried and tested method of checking the camouflage of positions or barriers from the enemy's point of view, but they should also now expand it to include the aerial dimension using their own drones.

SOFT-KILL SYSTEMS

So-called *soft-kill* solutions that **attack the communication or control units of micro-drones** appear to be an obvious solution at first glance. For this reason, such agents have become widespread in both the military and police sectors in recent years. Widespread jamming over a significant frequency range has several disadvantages, however. The **electromagnetic signature** of the emitter unit is enormous. There are other effectors on the battlefield, e.g. artillery, which, after reconnaissance and location determination, can destroy the jammer unhindered by interference from the jamming measures. In addition, we should not underestimate the **energy demand of such jamming systems.** Tracking the corresponding batteries or accumulators is a further burden on the company's own logistics. Installing them in vehicles can also fail because of interference with the vehicle's on-board communication network. Last but not least, **broadband jamming can cause severe restrictions on the use of one's own drone forces,** as the battle for Kiev impressively showed. In the spring of 2022, Russian forces, who had penetrated northern Ukraine from Belarus, had to suspend their own jamming measures multiple times in order to use drones themselves. The Ukrainian defenders used these time windows to launch their own drone attacks.

Solutions to the problem areas mentioned above are only partially available. It is possible for one person to carry jammers, like the HP47+ in service with the Bundeswehr, and use them independently of communication networks. Although they enable targeted use against individual drones without hindering the flight of one's own UAS, they require an advanced lead time and warning. Such systems also become quickly overwhelmed if they have to deal with and attack by multiple drones. Therefore, one should seek kinetic capability at the company and ideally platoon or squad level. This should comprise existing weapons with ideally minimal adaptation and modification.

HP-47+ Counter UAV Jammer: In use with the German Army since its 2016 deployment in Mali.

HARD-KILL SYSTEMS AS AN OPTION

Besides the jammer solutions that have been available for several years and which follow the soft-kill approach described above, hard-kill systems are currently increasingly becoming the focus of the discourse. **Kinetic effectors**—such as **nets fired from shotguns** or concentrated **water jets**—may be useful tools for police or local and static applications in the civilian sector. However, because of their very short range and rate of fire, they do not meet battlefield requirements.

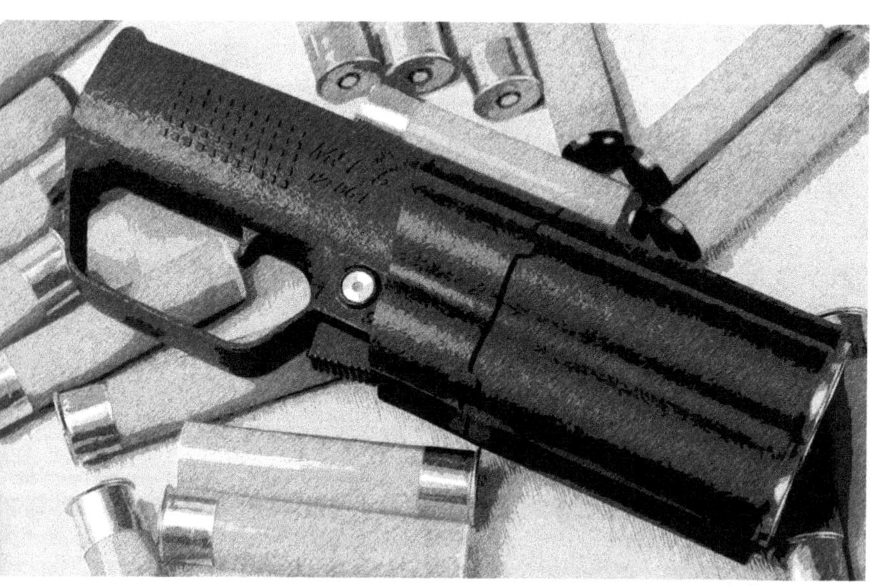

Signal pistol with special ammunition for close combat.

Shotgun cartridges make interesting hard-kill effectors, also place a low burden on the user—as engagements in Ukraine have shown. Users can also use 12-gauge signal flare launchers to defend themselves against FPV drones on their final approach. The launchers comprise compact and easy-to-use tube bundles that offer a multi-shot option and do not require sights. Increasing the probability of hits is to achieved by firing several shots and distributing the shot pellets. However, the actual combat value of such systems is more than doubtful.

Nonetheless, shotguns are appearing more and more frequently on the front lines of the Ukraine war and are enabling their operators on both sides of the conflict to make documented kills. What is striking, however, is that the ballistic performance of shotgun cartridges limits their use to short combat distances, which poses a massive risk, especially since small FPV reconnaissance drones can easily evade their effective firing range by adjusting their flight pattern.

In addition, shotguns, no matter how compact, as well as jammers, are separate weapons that are not an organic part of the overall equipment of soldiers and represent an additional burden on the battlefield. Despite this, both parties to the conflict have already demonstrated the successful use of shotguns to defend against small drones in the Ukraine war on multiple occasions.

Therefore, at the lowest level of drone defense, the goal must be to identify an effective means that many users have already introduced on a wide scale and ideally have already mastered operating it. This effector must be able to combat small UASs successfully and, ideally, also combat FPV drones or loitering munitionloitering munition, depending on their size and flight speed, at a distance of 100 to 500m. To increase the probability of a hit, we should plan for an inexpensive and fairly large supply of ammunition to create density of fire through a high rate of fire.

Small arms in caliber 7.62×51mm are clearly the most popular option here. The next step requires us to provide the identified effective means with ad-

Semi-automatic shotguns are being used against drones in Ukraine.

ditional capabilities for target tracking and projectile trajectory calculation. The high rate of fire mentioned would therefore put the focus on machine guns, but "Designated Marksman Rifles (DMRs) with their well-trained operators are also a conceivable option. This is precisely the conclusion reached by the British armed forces, which put out a tender for an effective weapon with exactly these requirements in autumn 2021. The aim was to procure a hard-kill system with the ability to combat NATO Class I rotary and fixed-wing UAS successfully at a distance of 20 to 500m on different flight paths.

One product that undoubtedly combines a large proportion of the capabilities mentioned is the **"SMASH" fire control sight family** developed by the Israeli company "Smart Shooter". In terms of dimensions, it corresponds to a night vision attachment, and integration into various small arms systems has already taken place. In the autumn and winter of 2020/21, for example, the Dutch armed forces successfully tested the "SMASH 2000 Plus" system on the Colt Canada C7 assault rifle. According to well-informed sources, this lead to a major procurement of the systems for current missions abroad. In addition, the Australian military performed successful tests, including firing DefendTex "Drone-40" loitering munition at distances of up to 100m. In the summer, the U.S. Army also granted final approval as part of the Counter-UAV program (C-UAV). The German Armed Forces have also tested the "SMASH" systems. In autumn 2021, the Bundeswehr published a video showing that they tested both “SMASH-AD” and “SMASH-X4” systems on the G95K (HK416 A7 in caliber 5.56×45 mm) and the G27 (HK417 in caliber 7.62×51mm).

The conclusion of the Bundeswehr's testing was that the systems would significantly improve the accuracy of hitting small unmanned targets within realistic scenarios. The system thus enables users to combat sUAS successfully with comparatively short training times. “SMASH” systems have now been procured as part of an “immediate deployment initiative”.

SMASH fire control system on an HK-417 assault rifle used by the Bundeswehr.

Another advantage of systems with highly automated image recognition is that they operate without emitting electromagnetic radiation. This means that the shooter's signature on the battlefield is very small, which has a positive effect on survivability. After detecting the target, the system calculates the projectile trajectory and only fires the shot when the necessary parameters are met, which also has a positive effect on signature and efficiency. Finally, even though there is a lot of automation, the shooter ultimately has the final responsibility for deciding whether or not to fire.

In these and other tests, it has been shown that small arms in 5.56×45mm caliber can successfully combat micro-drones within a range of up to 150m with a high probability of a hit. With larger aircraft at similar speeds, the range increases accordingly. The disadvantage of such systems on assault rifles during the tests was the low impact on the target with the lightweight construction of drones. A fixed-wing aircraft can be shot down with just one hit, but only if it directly hits the engine or critical electronic components.

On the other hand, fast quadcopters, such as those used for FPV carriers, are very difficult to impossible to hit with manually controlled small arms. **Remote-controlled weapon systems** solve this problem. Similar to a ground target mount, the weapon system allows the installation of machine guns of 7.62×51mm caliber with a belted ammunition supply. Because of the high density of fire and the stable weapon mount, it is easily possible to fight larger "Class I" drones at distances of up to half a kilometer. One can place such systems statically as a remote-controlled or (partially) autonomous weapon station or integrate them into an unmanned ground vehicle (UGV) to start the fight against enemy drones as a combat outpost without exposing one's own soldiers before one's own lines.

However, as the speed of UASs increases and their size decreases, machine-gun based systems quickly reach their limits of effectiveness. Another approach could be the use of **automatic grenade-launcher weapons**. For example, 40mm grenades offer enough payload capacity to bring down a sUAS through fragmentation. Programmable frangible ammunition, which allows detonation at a predetermined point on the flight path, is already available too. Therefore, it seems worthwhile to conduct tests against sUASs using such weapons platforms. Another advantage of such weapons is that they are already in widespread use and integrated onto many vehicle platforms—many of which are set up for remote operation.

CHALLENGE: LOITERING MUNITIONS

Defending against loitering munitions, especially those created by weaponizing small commercially available drones, is a particular challenge. Besides the aspects of detection and target tracking already discussed, which also apply here, their high cruising speed is worth mentioning. Depending on the model, their cruising speed increases by up to 70 percent on the final approach—making them even more difficult to intercept. Therefore, currently existing small arms cannot combat them, even with an increase in combat effectiveness through the addition of optronics intended for drone defense.

For this reason, soft-kill options through targeted jamming are currently coming back into the picture. In addition to GPS control, the telemetry transmissions required to guide loitering munitions is a starting point for jamming measures. However, when considering the last point, it should be noted that manufacturers of loitering munitions are increasingly using on-board artificial intelligence to locate, identify, and attack targets-making it no longer necessary to maintain a live telemetry transmission link. In addition, the moral principle of "human-in-the-loop" decision-making, which is predominant in the West, is not universally adhered to around the world. Some countries and non-state actors are comfortable with the idea of autonomous “killer robots”. In addition, targeted jamming of loitering munitions by portable effectors is fraught with the same problems as light kinetic systems because of the high flight speed of these aerial systems.

Another trend is the use of fiber optic cables for emission-free drone guidance. This technology, similar to wire-guided missiles, enables the significant reduction of the electromagnetic signature of the entire drone complex. Since the drone sends all data via the fiber optic cable, there is hardly any starting point for classic jamming. The fact that maneuverability is limited by the fiber optic cable gives both kinetic and passive measures increased importance in this case, for example, by pre-reconnaissance of approach directions and using these to align one's own defenses.

In terms of passive defense options, the analysis of the Ukraine war has shown that there is a great deal of overlap with measures against manned aircraft. The experience from Ukraine could lead to the conclusion that we can only successfully fight against reconnaissance drones. What sounds like a truism can—if taken further—lead to an effective approach. Both sides have been using **dummies and dummy positions** for some time, which can wear down the resource of loitering munitions, which are already scarce because of higher production costs. Furthermore, dummies and dummy positions also offer enormous advantages against enemy fire and, in fact, encourage early and ineffective firing, which also brings associated disadvantages. In addition, camouflage, particularly in the multispectral range, is of particular importance. Smokescreens and artificial (or natural) fog, in combination with dense, natural vegetation and additional artificial camouflage elements, can create great short-term camouflage and concealment of troop and vehicle movements—as well as static positions.

Vehicle-mounted metal net and grid structures occupy a special position in the study of protection techniques. Basically, the idea of (partially improvised) front armor and side plates on vehicles was first implemented in the Second World War. In the 21st century, the concept renewed as protection against shaped-charge anti-tank projectiles, such as the RPG-7, and later adapted further to protect the tops of vehicles against top-attack missiles such as the "Javelin". Further development of the design principle has led to its application in stationary contexts such as artillery positions, as well as in buildings and trenches. The forms range from simple improvised structures to expandable and modular factory-made products developed specifically for the purpose. The aim is to defeat the loitering munition on its final approach and, in the best case, to prevent its detonation. Russian manufacturers have also experimented with elastic materials or suspensions to deflect free falling warheads to the side. In practice, these net and grid solutions (often called “cope cages”) have not proven to be effective, as loitering munitions have still destroyed many vehicles with these structures. However, their spread continues unabated, and over time, targeted industrial development will need to prove in the future whether such measures can successfully protect based on empirical testing.

Finally, we should mention the importance of combating enemy reconnaissance systems, which is extremely relevant to the operational tactics of loitering munitions. Loitering munitions, by definition, combine the ability to reconnoiter, track, and attack targets with one device. However, loitering munitions have a significantly higher chance of success if persistent reconnaissance drones detect targets in the operational area and then transfer them to a loitering munitions operator for final fixing and attacking. As already mentioned, effectively combating the reconnaissance drones can also reduce the effectiveness of loitering munitions.

Another passive protection measure against drones is to build so-called 'Cope Cages' onto vehicles.

▸ CONCLUSION

Micro drones have become an indispensable part of the modern battlefield. Their use, and the performance parameters of the individual systems, will change in the future. However, their constant presence is guaranteed. Therefore, modern armed forces, as well as internal security authorities, cannot avoid establishing operational doctrines and the means to defend against them. This requires a holistic approach that combines passive and active measures and addresses challenges such as airspace management and deconflicting.

In the future, the technical and organizational consideration of small **hunting drones** will certainly also have to be examined. The conceptual consideration of such a means of action should begin today. We must also consider the constantly growing potential in cyber warfare capabilities. **Hacking and taking over enemy sUAS at the tactical level are certainly interesting approaches,** which could represent an important additional line of defense in the multi-layered defense concept in the future. In addition, a simple, cost-effective capability is required at the lowest tactical level in order to defend against micro-drones with a kinetic effector. At the company or battalion level, a network of additional kinetic, electromagnetic, and cyber effectors ideally supplements the combined approach of reconnaissance sensors for detection and target tracking to provide a shield over one's own forces.

Another approach is the targeted fight against the enemy drone forces before they can even intervene in the battle. For example, the infrastructure required for the operation of UAS, such as the provision of electrical charging capacity or repeaters to increase the range of the signal, can be a target of the company's own means of action. Furthermore, once we detect a sUAS, we must determine if we can locate the position of the operator and if the operator should then become a higher-priority target than the drone itself? After all, it takes much longer to replace a skilled UAS operator than it does to replace a drone.

Fundamentally though, the order of the day is to create the ability to defend against drones on the front lines of our forces and to establish this capability across the board.

DEFENSE IN THE AIR—THE DRONE-HUNTING DRONE

On April 3, 2024, the state coordination office for the Ukrainian defense industry, Brave1, announced via its social media channels that it was seeking constructive solutions for the development of their own fighter or interceptor drones. The framework conditions are open and public and allow anyone inside or outside Ukraine to submit proposals. "Brave1" is an interdepartmental coordination platform of the Ukrainian government that works across the Ministry of Defense, the Ministry of Economy, the Ministry of Digital Transformation and other government agencies. The aim is to ensure the coordination of private sector activities in the defense industry and to provide financial and organizational support as needed. Access is also explicitly open to foreign partners, whether through participation in "hackathons" or through financial donations.

The focus is on unmanned systems such as drones and unmanned ground and sea vehicles (UGV/USV), as well as cyberspace, intelligence and medical services, and the development of mine clearance technologies. The approach of making a call to the public is not new, as Brave1 has always encouraged interested parties to submit new ideas via its website.

The development aims to significantly limit the ability of the Russian armed forces to use unmanned aerial vehicles (UAS) for reconnaissance above the front or immediately behind it. The call from Brave1 explicitly named the Russian drone types and manufacturers Orlan, SuperCam, and ZALA. While the first two are reconnaissance platforms, the mention of ZALA can also imply that "Lancet" type loitering munition, produced by the Kalashnikov subsidiary, is also among their intended targets.

The concise technical requirements specified by Brave1 also support this assumption. According to the stated requirements, an air target with a speed between 100-150 km/h at a maximum altitude of 1500 m should be located, tracked and successfully intercepted or shot down. The requirement leaves open whether collision or detonation of the interceptor drone should destroy the target, or if an on-board weapon, which would allow the interceptor drone to be reused, should do

it. However, cost savings compared to conventional air defense systems, such as shoulder-launched anti-aircraft missile systems (MANPADS), is a declared aim of the project.

Intercepting drones with other drones is not a new idea either. In spring 2022, for example, the Israeli company Steadicopter presented a copter drone with an integrated "SMASH Dragon" fire control solution and the ability to carry tube launched weapons up to 40 mm caliber. However, the "Black Eagle 50E" manufactured by Steadicopter could not meet the required performance parameters published by Brave1.

New uses for old tools: the tripod and other support aids are once again becoming important in the fight against aerial targets such as drones.

Supplies for the front: FPV drones deliver aid packages.

THE KALASH-NIKOVS OF THE SKIES

3

MARKUS REISNER

DRONE WARFARE HAS BECOME ENORMOUSLY IMPORTANT—THAT IS ONE OF THE MOST SIGNIFICANT LESSONS FROM THE UKRAINE WAR.

This war is being fought by both sides with the use of drones of various kinds. Mini drones are playing an increasing role. The course of the war also reveals a dynamic of countermeasures and their neutralization or circumvention, which deserves attention because otherwise Western forces could fall behind. However, the war in Ukraine does not mark the beginning of the militarization of drones and the associated transformation of warfare.

The importance of drones for reconnaissance, tactical leadership and as a direct means of combat has been growing for over the past 20 years. This transformation of warfare actually began even earlier, but the Ukraine war is now the main catalyst accelerating these developments.

At the beginning of this century, drones were used on a large scale for military purposes, primarily by the USA, but also by Israel. The aim was to fight terrorists, militias, or irregular forces in remote and difficult-to-access regions without risking the loss of personnel or high-value systems. After September 11, 2001, there was a boom in drone warfare in the United States and other countries, both for reconnaissance and fire control, as well as for direct kinetic use. For example, the Obama administration relied heavily on drones to combat Islamist militias and terrorists in Afghanistan, Somalia, Yemen and Iraq, Pakistan and other countries during the Global War on Terror.

This development stimulated numerous scientific analyses that addressed different aspects:

- First, we examined the effectiveness of drone warfare in counterterrorism and asymmetric warfare. The key question here was whether and to what extent drone warfare could reduce the number of terrorist attacks against Western armed forces or allies, and how effectively the killing of leaders in terrorist organizations or Islamist militias could impair their ability to act. Another focus was the question of the extent to which drone warfare undermines the strategy of counterinsurgency by alienating the population that one actually wants to win over to one's own cause from the West.

- A very controversial debate also developed over whether targeted killing by increasingly automated, remotely operated systems did not exceed the limits of what is permissible under international law. In Germany, in particular, there was a very serious debate about whether drones should be bought at all.

- It was also pointed out that access to drone warfare technology would not remain limited to Western countries. As drones continue to spread, we can expect their use in asymmetric conflicts against Western forces and the opening up of new opportunities for terrorists, in particular.

- Other authors emphasized that drone warfare would also change the qualitative character of "regular" warfare between traditional armed forces. Drones would change the dynamics of the conflict between states and offer new options for escalation or delay.

- Today, it must be said that the last two points mentioned in particular hold high relevance, and many of the forecasts made at the time have proven to be correct. On the one hand, the use of drones is proliferating. We discuss this in the first section. The war in Ukraine has led to a dynamic in warfare with and against drones that has significant tactical, operational and strategic consequences. We will discuss this development in the following section. Finally, we will also discuss the likely strategic consequences of the growing proliferation and use of drones.

DRONE PROLIFERATION AND IRREGULAR WARFARE

Drones are now available in a wide variety of variants and are available to an increasing number of state and non-state actors. The rapid technological developments of the last ten years have made mini-drone systems (mostly of civilian origin) available to anyone and accessible for use anywhere. Both state actors and terrorist organizations have recognized these new possibilities since the mid-2010s. It was the “Islamic State” (ISIS) that first used commercially available mini drones on a large scale. Initially, these were primarily used to identify potential targets for moving car bombs (controlled by suicide bombers) (“Suicide Vehicle-Borne Improvised Explosive Device”, SVBIED).

However, ISIS also quickly developed even more innovative ideas. Their fighters were extremely successful in dropping small explosive devices from commercially available drones or causing mini drones loaded with explosives to crash onto targets in a kamikaze-like manner. During the battle for Mosul from October 2016 to July 2017, Iraqi security forces faced dozens of drone attacks every day. Not every explosive device hit its target, but random hits followed by spectacular explosions posed major challenges for the Iraqis and coalition forces.

ISIS produced the explosive devices it dropped according to its own quality standards and used a variety of different mini drones (mainly from the Chinese company DJI) in rotary and fixed-wing versions. The cameras it carried had a variety of tasks for reconnaissance, targeting and assignment, and the creation of useful propaganda images.

Word of these capabilities quickly spread, so that the use of easily obtainable drone systems increased in all conflict regions. From the Near and Middle East, these Tactics, Techniques, and Procedures (TTP) migrated to the Sahel region, Libya, and finally Ukraine. Many other terrorist organizations followed the example of ISIS using mini drones, including the Taliban, who now simply filmed their spectacular attacks using mini drones.

The year 2018 brought the first notable improvement in quality. In February, an (Iranian) "Saegheh-2" drone flew into Israeli airspace from Syria. They shot down the drone quickly, but the following attack on its ground control station resulted in the shooting down of an Israeli F-16. This was a turning point. The "Saegheh-2" that was shot down was strikingly similar to an American RQ-170 "Sentinel" reconnaissance drone. One example of this type disappeared in Iranian airspace in December 2011. After initial denials by America, then-U.S. President Obama finally called on Iran to return American "property". But by that time, Iran had apparently reproduced the American drone using reverse engineering. What made the incident in Israel really significant was the fact that this "Saegheh-2" had obviously been loaded with explosives. This was a nasty surprise for the Israelis, as the enemy—in this case the Iranian Revolutionary Guard—would have now been able to attack any place on Israeli soil. Israeli Prime Minister Netanyahu therefore presented the remains of the drone to great publicity (including at the Munich Security Conference).

Despite a strict sanctions regime, Iran has quickly risen to become a kind of "drone superpower", with the war in Yemen serving as a testing ground for the technology of unmanned systems. Despite the delivery and use of Western precision weapons and armed drones, an Arab coalition has not yet defeated the Houthis fighting in Yemen. The Houthis, on the other hand, could create a certain symmetry in the conflict by using various types of Iranian drones.

Although they could not counter the bombings of the Arab coalition forces, they could retaliate from hundreds of kilometers away. In 2017, the Houthis used "homemade" drones of the "Qasef-1" type for the first time. In appearance, these were clearly identifiable as the Iranian "Ababil-2" model. With a range of 150 kilometers and loaded with explosives, they already represented a potent weapon system. As a result, reports of their use increased. The Houthis claimed responsibility for several attacks on targets in Saudi Arabia and the United Arab Emirates. The attacks targeted critical infrastructure, i.e.

airports (including in Dubai) and oil facilities (pipelines and production facilities in Saudi Arabia). Some sources claim these attacks were carried out from over several hundred kilometers away. In April 2018, Saudi air defense systems shot down suspected drones at the airports of Abha and Jizan for the first time. Then, in July and August 2018, attacks with "Sammad-3" drones were allegedly made on the airports of Abu Dhabi and Dubai. In July 2018, another attack with a "Sammad-2" took place on the Saudi oil refinery in Riyadh. According to reports, the two types used have a range of up to 1,400 km. By mid-2022, the Houthi rebels in Yemen had fired over 430 ballistic missiles and 851 drones at the Kingdom of Saudi Arabia.

At the beginning of September 2020, a spokesperson for the Yemeni army, supported by the Houthi, announced in a press conference that it had once again attacked the Saudi airport in Abha with its own long-range drones. Reports like this clearly showed for the first time that drones and the effects of their use were no longer to be underestimated. A veritable drone war is still raging in Yemen and Saudi Arabia. While American MQ-9 "Reaper" drones hunt al-Qaeda fighters on Yemeni territory, Houthi are defending themselves against attacks by a Saudi-led coalition by using drones loaded with explosives. They mainly use drones supplied by Iran for this purpose.

This type of warfare repeatedly leads to spectacular successes. In September 2019, for example, the attackers launched an attack on the important oil production and distribution facilities of Khurais and Abqaiq in the middle of the Saudi desert. The experts described the consequences at the time as causing the "largest daily disruption to oil supplies in human history. The outage caused the total loss of supply to Saudi facilities, including around 5.7 million barrels of oil production per day——more than half of Saudi Arabia's recent production and around six percent of global supply——as well as two billion cubic feet of gas production per day. Attacks like these show that drone warfare is now an integral part of every conflict zone. Drones are not just the preserve of powerful state actors (e.g. the USA, Israel, Great Britain and France), but are increasingly being used primarily by non-state actors.

Today, the phenomenon of drone attacks by non-state actors in the conflict region of the Middle East is nothing new. Events in the conflict zones in Iraq, Syria, Yemen and the Levant (i.e. Israel against its multitude of enemies) have been full of reports of drone attacks for several years. These range from the use of improvised armed mini drones to unmanned systems the size of small aircraft. As early as 2004, Israeli soldiers made an unpleasant discovery: the terrorist organization Hezbollah had apparently used mini drones for reconnaissance. Over the next 24 months, this capability was further expanded, and in 2006, the next surprise came: Hezbollah fighters attempted to use drones equipped with explosives in targeted attacks against Israeli soldiers. Their systems are also constantly being further developed. Hamas, which has had drones since 2021, also used them as an important part of their attack on Israel on October 7, 2023. Hamas used its drones to destroy surveillance towers and IDF "Merkava" main battle tanks.

In Ukraine, reports of mini drones flown by the so-called separatists increased from 2014. The analysis of the models used confirmed that the drones were Russian army models, not manufactured in Luhansk and Donetsk. In Syria, regime troops and rebel groups also copied the operational drone tactics of ISIS.

And the country became known for a further improvement in the quality of operational management: From January 2018, entire swarms of drones repeatedly attacked the Russian Khmeimim Air Base and damaged or destroyed several Russian fighter planes. There were indications that the attacker had directed the individual drones to the target using a guidance beam—a skill whose complexity cannot be attributed to the Syrian rebels. The perpetrator of this attack, which was carried out over a long distance, remains a mystery.

However, the fact is that the operations of the Russian Air Force contributed significantly to the success of the Syrian armed forces, meaning that disrupting these operations was in the interests of numerous actors. It has therefore become clear that the use of mini drones is of interest not only to non-state actors but also to states that do not want to be associated with an attack. The drone perfectly suits this purpose. The absence of markings, and most importantly, a human pilot, in the drone's remains leaves room for speculation about the originator of the operation. Even if the technical design indicates a certain country of origin, proving that this country launched the attack is very difficult.

The more complex capabilities a drone has, the more technical effort is required to build and use it. People can order simple systems online, but larger models come from military research and production. For example, since 2014, the Ukrainian armed forces and their volunteer units have repeatedly shot down and captured Russian “Forpost” and “Orlan-10” models in eastern Ukraine. An analysis of the technical capabilities has shown that these systems are suitable for much more than just reconnaissance. They enable real-time target assignment for artillery systems with different ranges (e.g. TOS-1 or BM-21, BM-27 multiple rocket launchers or 2S19 self-propelled howitzers). Interestingly, the Russian “Forpost” is a further development of the Israeli IAI “Searcher”, which was developed by Israel in the 1980s and sold very well on the export market. Drones of the “Orlan-10” type have devices that enable control of GSM radio signals. Russian armed forces use the

"Leer-3" system in this configuration. Orlan-10 drones are clearly Russian developments and were first delivered to the Russian armed forces in 2013.

The list of drones of different types and sizes that different actors have successfully used goes on and on. It is noteworthy, however, that armed drones are no longer used only by countries known for their drone warfare, such as the USA or Great Britain, but also by countries such as Iraq, Nigeria and Iran. China has recognized the gap in the global arms industry and is already supplying systems to order that are comparable in size and performance to American MQ-1 "Predator" and MQ-9 "Reaper" unmanned aerial vehicles (UAVs). Arab states are also proving to be eager Chinese customers. As a result, sightings of wrecks of Chinese drones (e.g. "Wing Loong") or Turkish models (e.g. "Bayraktar TB2") are increasing in Libya and Yemen. Drone warfare is therefore no longer the preserve of the well-known actors.

What is special about the war in Ukraine is the mass use of small drones.

DRONE WARFARE IN THE UKRAINE WAR

Russia's renewed invasion of Ukraine on February 24, 2022, sparked a real drone war with both the Russians and Ukrainians making widespread use of drones on a large scale. In the early stages of the war, Ukraine benefited more from the use of drones than Russia. It already had some powerful drone models in February 2022 when the war began, and then later also received donations and deliveries of other systems, such as the Quantix "Recon" and "Vector".

Among other types, Ukraine also acquired the Turkish model "Bayraktar TB2". This drone is characterized by its small, compact design and the ability to carry air-to-ground missiles. "Bayraktar TB2" can launch missiles of the MAM-L, MAM-C and MAM-T ("Mini Akilli Mühimmat"—Smart Micro Munition for Drones) types. In 2020, Ukraine used these Turkish drones for the first time to monitor radar stations in Crimea. A year later, the TB2 was used to fire on an artillery position in the separatist areas of eastern Ukraine. Since the start of the war on February 24, 2022, these drones have also been hitting targets on the Russian side of the border. All in all, Ukraine has conducted many successful attacks on Russian forces with "Bayraktar" drones.

By the end of April 2022, TB2s were responsible for the destruction of almost half of all disabled Russian surface-to-air missile systems (e.g. of the “Buk M1” type). They also destroyed at least six armored vehicles and five towed artillery pieces. Three TB2 drones were also involved in the sinking of the Russian cruiser “Moskva” in the Black Sea. They disrupted the defense system of the Russian cruiser “Moskva,” located the ship, and thus enabled the successful attack on it using two Ukrainian-made “Neptune” missiles. Sea drones have also played a role in both the attacks on Russian military facilities in Crimea and the attacks on the Kerch Bridge.

Ukraine also has loitering munitions (“kamikaze drones”), such as the Polish model “Warmate”. This drone plunges into a target and destroys it upon impact. The warhead allows it to penetrate even strong armor. In addition, there are deliveries of models from NATO and the USA. Specifically, these are the “Switchblade” 300 or 600 and “Phoenix Ghost” systems. These systems can destroy tanks with their shaped charge warheads and thus have a significant effect on the battlefield.

In response to Russia's strategic air strikes, Ukraine has also repeatedly carried out attacks on Russian territory, for example, with Tu-141 drones. The Ukrainians were also the first to use the first-person view drone (FPV). This technological innovation originated in the hobby drone scene, where enthusiasts developed drones that are to VR glasses and controlled as if the user were sitting in the drone like a pilot. The Ukrainians use these drones and its control system, but weaponize them with explosive payloads, such as the warhead of an anti-tank missile, and then fly them directly into a bunker or a vehicle. Ukraine has used this method successfully on the battlefield and continues to deploy devices, but only on a small scale because it no longer has the extensive military-industrial capacities to support their mass production.

Russia initially attached little importance to drone warfare. However, this changed rapidly over the course of the war. Its armed forces now have various drones of different weight and performance classes, such as tactical and smaller drones (e.g. "Forpost" or "Orlan-10" and "Orlan-30"), but also larger models (e.g. "Orion", comparable to the "Predator" or "Reaper" drones of the U.S. armed forces). Russia also uses kamikaze drones, such as the "Kub-Bla" model in Kyiv and the "Lancet" model in Donbass. Drone reconnaissance combined with artillery fire has proven particularly effective for Russia. Unlike Ukraine, Russia has been very successful in using drones for strategic air strikes on Ukraine's critical infrastructure.

The Russian armed forces did not initially have the necessary numbers of drones available but were able to import many of them from Iran from summer 2022 ("Shahed-131/36" drones). Iran has so far supplied about 3,000 units to the Russian military.

This means that Russia has a cheap, efficient weapon system that puts high pressure on the Ukrainian defenders—as seen from the fact that Ukraine has so far shot down over 2,000 "Shahed" drones. However, it is worth mentioning here that the "Shahed-131/136" primarily seeks to engage the Ukrainian air defenses and deplete their stock of anti-aircraft ammunition, with the aim of creating openings in the Ukrainian air defenses for targeted cruise missile attacks. Moreover, a "Shahed" missile costs around 20,000 euros, while each "IRIS-T" anti-aircraft missile in Ukraine costs a whopping 450,000 euros. Russia would not have been able to carry out its strategic air campaign without the "Shahed" drones. Ukraine, on the other hand, has only been able to carry out symbolic attacks on targets on the Russian hinterland and Moscow with low-performance long-range drones.

Russia is now equipping its armed forces with drones on a massive scale. "Shahed" drones are now assembled in Russia and increasingly equipped with Russian components. Russia has also stepped up its use of FPV drones, with drones also supporting it in electronic and information warfare. For example, the Russian "Orlan-10" drone was part of an information warfare campaign that transmitted targeted messages to Ukrainian soldiers as part of psychological warfare efforts.

To date, Ukrainian and Russian armed forces have deployed thousands of drones in confined spaces and simultaneously on the battlefield. Reconnaissance drones create a transparent picture of the situation, while attack drones have a destructive effect. In particular, FPV drones, which can be produced cheaply and equipped with shaped charge explosive devices, allow the operator to effectively attack anything that moves on the battlefield from a safe distance.

The use of drones in Ukraine has created an operational dynamic since February 2022 that has led many observers to speak of a "revolution in warfare" or at least a significant qualitative change. However, not everyone shares this view. In fact, it seems to depend primarily on the specific circumstances of drone use, i.e. there is debate among observers on whether and how the specific course of the war will change in the long run in the struggle between drone use and drone countermeasures, as well as the roles that the industrial capacity and financial endurance of both sides play.

If we take the two factors of "quality competition" and "industrial power" into account, the current picture is not in Ukraine's favor. Improved electronic countermeasures by the Russian side have significantly reduced the effectiveness of Ukrainian drones. Russia has also disrupted about three quarters of the Ukrainian drone attacks, i.e. diverted and neutralized them as a precise means of attack (jamming). As a result, the Ukrainians can no longer use the FPV drone system they invented as effectively as in previous months. Long-range Ukrainian drones are also being jammed increasingly effectively by the Russian side. As the Ukrainian Chief of General Staff Valerii Zaluzhnyi pointed out in an article for the "Economist", his country could only resume the offensive to recapture the occupied territories once there had been progress in electronic warfare (meaning defense against Russian jamming) and other weapons systems.

In addition, the production of FPV drones in Ukraine has become increasingly difficult since China banned the export of the civilian drones required for this purpose to Ukraine. Additionally, the factor of industrial power should be considered. Russia is already outpacing Ukraine in both the quantity and quality of drone production-for example, Russia has announced that it plans to produce up to 6,000 "Shahed" ("Geran" in Russian) drones in the near future. Russian air strikes have also severely diminished Ukraine's defense industry capacity. Russia is now flooding the battlefield with its drone systems and also jamming those of Ukraine at the same time. The Ukrainians are successfully jamming the Russian drones too, but their capacity is more limited.

The effects of these developments are foreseeable. Ukraine's operational command has always been successful when it was mobile, but the Russians have forced them into a stationary war of attrition. The war is thus becoming a meat grinder that benefits the Russian side. Only in this way can Russia exploit its great capabilities, for example, in the massed use of artillery. Ukraine wanted to break this dilemma with its offensive, which began on June 4, 2023. After the offensive failed to achieve its objectives, Russia forced the Ukrainians back into trench warfare. Both sides are now in a kind of stalemate. And technological developments are reinforcing this trend. While Ukraine has introduced very innovative new weapons systems in recent months, the Russians have copied these innovations and started producing them themselves.

In the meantime, we have the so-called 'glass battlefield' of Ukraine, on which no tactical or operational maneuver is possible without detection for the attackers or defenders. Attack drones or drone-controlled artillery swiftly detect and crush any movement by mechanized forces in its tracks. In addition, there are mines and ambushes using anti-tank guided missiles. Both warring parties are being forced into a miserable war of attrition and unless there is a quantitative increase in Western military aid, this will be to the detriment of Ukraine.

OUTLOOK

The performance of drone operations in war zones over recent years makes us aware that unmanned weapon systems have become one of the first weapons of choice in modern warfare. All the developments described above show that state and non-state actors are increasingly relying on drones in current conflict zones. Drone warfare is clearly a phenomenon of modern warfare.

Just as Western states are emphasizing the ability to wage war more precisely, and possibly more humanely, other actors are also discovering the benefits that the use of drones can bring them. Drones are a cheap and efficient tool and, if used correctly, can have a strategic effect far greater than their cost. Above all, they have two major advantages: Firstly; the actor using drones does not have to worry about human pilots. Secondly, the actor using drones can always publicly deny the origin of wreckage. Therefore, we can assume that in the future, the world's war zones will see an increase in the appearance of more flying objects of "unknown" origin.

It is also only a matter of time before the first drone controlled by terrorists heads for a football stadium or critical infrastructure in supposedly safe countries——with criminal intent and with devastating effects. As shown, we can already see that drones can serve as weapons carriers, easily adapted by terrorists to carry air-to-ground weapons or explosives. Similarly, drones could also carry chemical or biological weapons. If such a deployment were to take place in swarm form, it could actually have catastrophic consequences.

The article was published in 'SIRIUS. Zeitschrift für strategische Analysen', issue 1/2024. Reprinted with the kind permission of the editors and the publisher de Gruyter:

www.degruyter.com/journal/key/sirius/html

What is special about the war in Ukraine is the mass use of small drones.

FUNDRAISING FOR FPV DRONES

HELP HEROES OF
UKRAINE

The “armament of the people”—weaponized FPV drone systems are often financed through public crowdfunding.

4 EXPERIENCES FROM UKRAINE

PROTECTION AGAINST THE DRONE THREAT

GUSTAV FREIMANN

In the next section, I will share my experience of defending and protecting against drones in the Ukraine War.

To help readers understand where this experience comes from, I will briefly introduce myself and give an insight into my background. I served in the German Bundeswehr for twelve years as an engineer sergeant, mainly in construction and tank engineering. In addition, I took part in several foreign missions during my service. After my service, I completed studies in the fields of security and management. In my subsequent civilian career, I spent several years in these areas in low and middle management.

After the Russian invasion of Ukraine on February 24, 2022, I decided to not simply accept such a thing. I believed I could be of most help on the ground with the military knowledge and skills I had gained over the years. That's why I went to Ukraine at the beginning of May 2022, joined the Ukrainian armed forces and since then—as of February 2024—except for a few weeks of vacation, I have been working as an officially recognized combatant.

In the first few months after my arrival, I worked as an instructor in the Ukrainian army, but then, because of my range of skills, they assigned me to a special unit of the International Legion. This unit comprised former members of various special forces and specialists from various disciplines. The range of tasks included "Recon, Ambush, Harassment, Attack of HVT and Targets of Opportunity". I primarily acted as a combat engineer and "AT" ("anti-tank") gunner. After almost a year, I moved to an EOD ("Explosive Ordnance Disposal") unit because that's where my skills are best suited.

During this time, I served in the Odessa/Mykolaiv, Kharkiv and Luhansk Oblast, and Zaporizhzhia Oblast areas. I took part in the Kharkiv Offensive, the Battle of Bakhmut and the Spring Offensive in the South, and more.

DEFENSE AGAINST DRONES

I had my first contact with drones during my time in the German army. Primarily positive, for example, when our own reconnaissance used models such as "Mikado", "Luna" or "Aladin". But also negative when the opposing side used drones. I encountered this particularly frequently during my deployment in Iraq as part of the operation against the "Islamic State" (ISIS).

The first incident occurred when a U.S. Army firebase received astonishingly precise counterfire. A U.S. artilleryman suffered serious injuries and died while we were treating him in our camp. The ISIS fighters had captured several guns from the Iraqi government, but had only a few well-trained artillerymen and little ammunition, which is why the fire was usually sporadic and rarely effective. People later suspected that the COTS drones ("Civilian off-the-shelf") used by ISIS in this area were not only employed for reconnaissance but also for target observation and fire correction, resulting in the accomplishment of such precise strikes.

But not only that, they also employed drones as weapons. They loaded COTS fixed-wing drones with explosives and guided them into the positions of the allied Kurdish fighters: these were quasi-improvised loitering munitions or forerunners of today's FPV drones. In addition, there were reports from reconnaissance and the Kurds that ISIS was also loading these attack drones with chemical warfare agents that it had improvised in occupied Mosul, mostly chlorine compounds or mustards such as S-mustard gas based on sulfur compounds. Since we had no means of defense, protection was the main priority. We always carried complete personal NBC protective equipment and wore it when we received a drone alert. We also sought whatever protective structures were available. Fortunately, there were no direct attacks, but spending some time in full protective gear at 45 degrees outside also severely affected performance.

In Ukraine, however, my experience with drones increased exponentially. Right from the start, the omnipresent use in all areas was noticeable. In army units, drones were often already present at platoon level, and additional ones in specialized teams at battalion level. During the operations on the front, there were always encounters with drones. We quickly realized that we could no longer carry out infantry operations as before under these circumstances and that we needed to make adjustments.

We achieved significant progress during my time in the special unit. During my time in the special unit, we made significant progress, particularly in identifying quadcopters as the primary threat, as they searched for individual infantry troops. Larger drones, especially fixed-wing aircraft, mainly looked for vehicles, equipment or larger units. Reconnaissance by quadcopters could have several consequences. For example, if one's own position was compromised, the consequences might not be immediately clear, but it could result in immediate artillery fire or 'drone drops'. Mortar fire usually followed the artillery, usually of 82 mm or 120 mm caliber, and in more unpleasant cases, even from howitzers or tanks. 'Drone drops' is a colloquial term for dropping grenades or improvised explosive devices from a drone, but these are easy to spot though because the drone has to hover over the target in order to aim accurately.

In our unit, we used drones at squad level and in specialized teams, and we started to include them in almost every exercise. This had the advantage of training each other because the operators learned how to spot infantry squads and how to best approach them, and we infantrymen learned how to best protect ourselves from the drones. A positive byproduct of this was that it created video footage, which could be evaluated afterwards to improve performance.

Key points that were identified during these exercises and verified in actual missions:

1. DISPERSAL

The greater the distances between the individual soldiers, the more difficult it is for the operator to see how large the unit is, or even to read an intention. Furthermore, it minimizes the potential damage from enemy fires or drone drops. In the Bundeswehr, the rule was eight paces distance, roughly four to five meters, between soldiers. We now regularly used eight to twelve meters distance between soldiers. In addition, we sometimes use further subdivisions. For example, a small team of two to four soldiers often goes 20 to 40 meters ahead of the main part of the squad, to scout the way and prepare it (especially under the threat of mines) if necessary, or also to conceal the actual full size of the squad from enemy reconnaissance.

2. CAMOUFLAGE

Optical camouflage works very well against drone reconnaissance. As a rule, a uniform in a good camouflage pattern, combined with appropriate tactical behavior, is sufficient. In our team, the most common camouflage patterns were MultiCam, British MTP, Ukrainian M14, German Flecktarn and French CCE. MultiCam and MTP were particularly good here, as they worked in most environments. CCE and Flecktarn were outstanding in wooded terrain, especially when it was blooming, but were not very helpful in other environments. Similarly, the Ukrainian M14 was good for fields and meadows, but otherwise not particularly impressive.

Two factors that are noticeable for drones are reflections and movements. When we talk about reflections, we are not just talking about materials, but also about skin. We have therefore adapted our uniforms accordingly. No short sleeves, even in midsummer, and also tube scarves, balaclavas or veils for the face, as these are more comfortable and effective in the long run than camouflage make-up. When moving, fast and hectic movements are more noticeable than slow and methodical movements.

When you come into contact with a drone, it is therefore a mistake to run immediately for cover or to your fighting positions. Our SOPs (standard operating procedures) when there was visual or acoustic contact with a drone were to stay in position, do not look up, crouch down or lie down as slowly as possible, and only take cover, such as under a bush, if very close by. However, since it is difficult to tell which drone belongs to whom, this was the drone precaution for every contact. As drones were very prevalent in the operating area, a tactical movement (such as an approach march or patrol) that would normally take an hour could now take three hours. Operational planning must now consider this additional time for movements. In the same vein, when possible, operational planning should select areas with good overhead cover, such as dense forest areas or paths through buildings.

Another aspect of camouflage is the heat signature, since drones with thermal optics are now in widespread use. The most important thing here is to create visual separation, as such optics can only read the heat from the closest surfaces. This can be, for example, a wall, a canopy of leaves, a stretched tent or tarpaulin or even window glass. Placing this visual protection element at a sufficient distance from the nearest heat source is crucial to prevent heat buildup and ensure that the drone perceives it as the ambient thermal signature. Hardly possible on the march, but certainly workable in fixed positions or buildings.

3. PASSIVE DEFENSE

It is extremely difficult to shoot down a drone with small arms. Attempting to do so can also reveal your own position to the drone or nearby enemy forces. Therefore, you should always reserve this as a last resort against drones in very close quarters.

From the beginning of 2023, the use of loitering munition and FPV attack drones also escalated. Models such as the Russian "Lancet" in particular primarily seek large targets such as vehicles. We countered this essentially by adapting the types of movement and when stationary. Vehicles rarely or never stay on the front line of battle. Marching speeds were increased to between 70 and 140 km/h. Stationary times, e.g. for unloading, mounting or dismounting, were limited to a few minutes—— rarely more than three. VDOs ("Vehicle Drop-Off", location of dismounting) are between two and nine kilometers from the target object, depending on the danger situation, with the rest of the distance being covered on foot, and using cover. When not moving, one of the essential elements is cover from above. On the one hand, to avoid reconnaissance, and on the other, to increase the chance of a drone failing at this layer in the event of an attack.

COTS FPV drones are relatively fragile and a collision with branches, for example, can destroy rotors and thus affect the flight path. This may also trigger the sensitive ignition mechanisms of the improvised explosive payload. A common practice is to set up

FPV drones with explosive devices have long become commonplace weapons in the Ukraine War.

catch nets made of wire mesh or to use bird nets to increase the chance of catching FPV drones, loitering munitions or dropped explosive devices—sometimes also working to trigger their payloads early. This is particularly useful in static positions such as field fortifications, foxholes, or command posts. Because of the significant rise in using FPV drones against infantry targets, it is necessary to take these measures into account in this context.

Another aspect is tactical behavior in rest areas, troop housing, or command posts away from the front. Here, the danger comes less from the drone as an effector, but more from reconnaissance. High-flying fixed-wing models in particular—such as the Russian “Orlan”—can reconnoiter up to several dozen kilometers behind the front. Targets identified are then usually attacked by Russian rocket artillery. It is therefore important to keep the signature low in such areas too. As an example, soldiers park vehicles several hundred meters or even one to two kilometers away from the accommodation, and they avoid storing any military equipment outdoors around the accommodation.

4. ACTIVE DEFENSE

While protection against drones can be easily improvised using camouflage, tactics, or even construction measures, in the area of defense, your reliance is more on the availability of material and equipment. You need effectors, ballistic or electronic, to ultimately shoot down the drone, and ideally also means of detection.

As already mentioned, it is extremely difficult to shoot down drones with infantry small arms. On the one hand, you only see and hear quadcopters very late, and then they are very difficult to hit because of their small size. Drone guns have proven to be helpful here. These are jammers, i.e. Jammers, which are aimed like a rifle, have also proven to be helpful. The problem is that they usually have to act on the drone for several seconds to shoot it down, so they are best used when it is hovering. Therefore, when one of these devices was available to support it, we usually positioned it in the last third of the formation. This meant that we could take advantage of it when a drone took a closer look at the front positions and hovered over them.

Our own operator always had a commercially available spectrum analyzer with him for detection. This shows which radio frequencies are being used in the current area. This makes it clear whether the area is being jammed by the other side or whether something is in the frequency range for drones (usually between 2 and 5 GHz). Some units have even combined such systems with directional antennas to determine the direction from which a signal came.

FPV drones pose a major problem. Because of their small size and high speed (up to 160 km/h), you hear them more often than you see them. And even then, it is usually not until it is already too late. Shooting them down with small arms or drone guns is almost impossible. Jammers, which block an entire area, have proven to be the most effective against this. Static or portable models exist, although there are far too few available. Both sides are therefore trying to retrofit them with commercially available material. Some Russian tank units equip almost every vehicle with a jammer. However, the use of such jammers also tells a trained operator that there is something worth protecting in the area that he can no longer fly to. This information can then help derive options for action, such as closer ground reconnaissance.

Electronic warfare is also a major means of combat operations. The Russian side uses it immensely and still has the upper hand in this area (as of February 2024). This affects Ukraine's entire C4 (Command, Control, Communications, Computers) capabilities, but is also responsible for most of the material losses in our drone teams.

Most of the successful shooting down of fixed-wing reconnaissance aircraft such as “Orlans” that I could observe were because of MANPADS or 9K33 “Osa” anti-aircraft missiles. However, since we had limited resources, we always had to consider whether the potential reconnaissance target was valuable enough to justify using such expensive weapons.

A completely different facet was the use of loitering munitions against infrastructure targets and to terrorize the civilian population—primarily with the Iranian "Shaheds". Since our unit received logistical and material support from the Odessa and Mykolaiv areas, I was in these cities more often in 2022, especially on the day when "Shaheds" were deployed for the first time. In short, it was chaotic. Throughout the city, we could hear small arms fire, and until the situation became clear, we assumed it was a coordinated multiple attack by Russian infiltrator teams. What we were actually hearing, however, were the desperate attempts of various guard and police units to fend off the "Shaheds" with what they had. Later, they found an adequate and efficient means of combating the "Shaheds" through the German anti-aircraft tank "Gepard"—but I find the actions taken in the meantime interesting.

On the one hand, besides the regular anti-aircraft units, they formed ad hoc units that brought together available personnel and material from various branches of the armed forces to defend against the "Shahed". Equipped with machine guns, they stationed themselves along the potential flight paths and fired at the targets in a coordinated manner—sometimes with dozens of weapons at the same time. This worked particularly well at night through the help of searchlights. This was partly because the "Shaheds" did not recognize when they were being illuminated and thus did not perform evasive maneuvers, and partly because it made it easier to concentrate fire. The Air Force also intercepted

"Shaheds" from above in open airspace whenever possible as well.

5. ALTERNATIVE DEFENSE

The drone fight in Ukraine is ground-breaking in terms of scale, but still relatively backward in terms of technology. There have already been deployments of autonomous systems in the past, such as the Turkish "Kargu-2" in Libya in 2020. However, if the drone is not autonomous, this means that not only the drone is a potential counter-target, but also its operator.

Hunting enemy operators was a primary task of our drone teams. Our drone teams divided this into two tasks: finding and fighting enemy operators. Finding enemy drone teams is not easy, as they are small and always operate from cover. The normal search using a quadcopter is therefore most likely to be successful if you can assign a specific area as the starting point of a radio frequency in advance. Inspecting crashed enemy drones can also provide clues as to the position of its operator. Alternatively, stalking, i.e. following an enemy drone using your own drone, is also quite successful. The fight then involves using classic methods such as drone dropping, FPV use, or artillery fire control, but it also includes placing IEDs by drone, especially at regularly used positions.

At the end of 2022, they assigned me for a few weeks to support our mortar team, which had a shortage of personnel. Here I worked as a loader and gunner on a rather old 82mm mortar from the 1950s. Despite its age, we could fire quickly and precisely because we used the Ukrainian fire control app "Kropyva". Things got interesting when we worked with a team that had a civilian drone detection system. This essentially hacked and read the data traffic of enemy drones. Although the system could only detect DJI drones, around 70% of the total drones in use at the time were from this manufacturer—so this worked well.

When we used this device, we rarely spotted fewer than 80 drones at the same time on a front of around ten kilometers on a regular day. Now we made use of a specific function, the recovery point. This is the point to which the drone will fly back independently. If you don't change the settings on some drones, many drones automatically save the point where they started as their return point. Professional operators "jailbreak" their drones—this is what they call overcoming the usage restrictions and changing the software—and often remove this function altogether. Inexperienced operators are usually not even aware of this function. Such users also do not change their position often. We could see these points on the system and transfer them to our fire control app. On some days, it took just under three minutes from the discovery of such a point to the impact of our first shell on it.

A completely different approach, which was mainly used by the Russian side, is area denial, the denial of potential areas of operation. Even if drones can operate from several kilometers away, a drone team must first be within this range and secondly take up a covered position. If these factors are prevented by firing or laying mines, this significantly reduces the efficiency and survivability of the drone team.

I encountered this particularly in the fighting in the Kharkiv Oblast before the Kharkiv offensive. This was largely determined by the mixed landscape of fields, villages, and forests. We regularly used the forests to disrupt the occupiers in the next village, for example, with snipers or drone attacks. The Russians quickly realized this and tried to make these forests impassable. To do this, they either gathered their vehicles, such as battle tanks and armored personnel carriers, as well as mortars, grenade launchers and machine guns, at irregular intervals and on suspicion sprayed a suspicious forest area for several minutes (this is the classic 'recon by fire' tactic developed during the Rhodesian war). These unpredictable fire attacks actually prevented us from using some areas. Today, FPV drones make it easy to locate and attack concentrations of vehicles, so this tactic is rarely used. Mining, on the other hand, creates an even longer-term blockade of potential positions, either by manually laid anti-personnel mines or often by artillery-launched grenade mines, especially in areas that were inaccessible to the Russians. For example, in some forests, the Russians have used Grad rocket launchers to lay so many PFM-1 anti-personnel mines that the area was no longer passable.

OUTLOOK

Drone warfare has a well-established history and drone usage will continuously increase and evolve to play an indispensable role on the battlefields of the future—not only as a threat for one's own soldiers but also as a benefit. The development of new weapons and the protection and defense against them are constants that run antagonistically like sine waves through the history of warfare, tactics and technology. We have made much progress with drones as weapons, and now it is time to reach the same level in terms of defense. Without this, experience shows that things will be extremely unpleasant.

EFFECTIVELY DISRUPTED—THIS IS HOW JAMMING WORKS

Explaining the way a jammer works is simple: A signal generator typically produces a "white noise" that resembles that of an analogue radio when reception is lost. This generated signal can take on various amplitude forms (sawtooth, sine, etc.), and usually has to be amplified by a high-frequency amplifier, which is then transmitted through an antennal. In order to disrupt a frequency, the jamming signal usually has to be stronger or more powerful than the information signal to be disrupted.

Let's take the example of a drone: the drone communicates with the remote control, and depending on the output power of the remote control and the drone, the range can be several kilometers. The environmental conditions are not insignificant though. If you want to block the signal from the remote control to the drone, the jammer has to send out a stronger or "louder" signal. The situation can be compared to communication between two people: if a third person (the jammer) comes along and shouts louder than the two people talking, then their communication is disrupted. Ultimately, the signal superimposes itself and overrides the receiver, disabling its ability to assign and filter out its actual information signal. It can then no longer assign and filter out its actual information signal.

Locating the drone is much more difficult. There are several approaches to locating a drone: radar, frequency triangulation, camera, or noise analysis. But so far, there has been no system that could locate a drone or the drone pilot with 100% accuracy or reliability. Several components would have to work together to locate a drone reliably, and the data from the systems would have to be evaluated accordingly and ultimately verified. The number of false reports of drone detection has so far been quite high. Challenges include; drones are of different sizes—the smaller the drone, the more difficult it is to detect visually; different drones also use different frequency ranges—and if these remain undetected, the drone can enter airspace unhindered. If frequency monitoring is successful, the drone pilot can be roughly located by triangulation. So far though, there has been

no system that could locate a drone or the drone pilot with 100% accuracy or reliability.

In Ukraine, for example, detectors seem to simply "listen" to the publicly available frequencies and emit an alarm tone when certain signal levels are exceeded. Soldiers often carry commercially available, Chinese-made omni-directional jammers with low output power. These jammers can be sufficient for drones in private use, but they become ineffective for drones that have been professionally modified.

Professional jamming systems work on several frequencies at the same time and are flexibly programmable. There are also directional jammers that can transmit in a targeted direction and thus reliably disrupt drones even at great distances. The sooner a drone is jammed and disrupted, the better. You never know what danger the drone poses (reconnaissance, attack with explosive devices, or possibly a kamikaze attack).

There are currently few means available for intercepting so-called kamikaze drones. These usually work using a camera image and fly to a target as precisely as possible under the control of the operator. Other drones have the capability to be pre-programmed and do not require remote control.

THE TEN COMMAND-MENTS OF DRONE DEFENSE

5

1. THE HEAVENS ARE NOT YOUR FRIEND

Heavenly forces may be on your side, but the sky itself is not your friend. Aerial observation is always possible. Most times, you won't be able to tell if you're being seen, so you can assume it's happening most of the time. The "glass battlefield" results from mass drone use. And even if you can't see or hear the drone, it can still observe you. Then there are drone bombers that drop bombs, and now FPV kamikaze drones that attack you personally.

2. LEND ME AN EAR

Different types of drones sound different. If you listen carefully, you have a chance of recognizing whether there is something there sooner than with your eyes. If the drone is above you, it definitely works; the further away, the more difficult. It may also have something to do with the size of the aircraft. The most difficult to hear are approaching FPV drones; you only recognize their high-pitched tone shortly before they reach their destination. By the way, learning also means practicing here—and playing with flying drones so that you gain experience before things get serious.

3. LOOK CLOSELY

The second way of understanding is that we understand what the drone sees. Learning this is relatively easy and impressively open in Ukraine, where they document everything and every situation with videos—from observation to attack. So: watch videos of operations! And think: What does the drone see? How is it used? And how can we behave to prevent being seen or attacked?

4. CAMOUFLAGE IS HALF THE BATTLE

Camouflage is not a special topic, it should always be a matter of course. However, with the increasing number of drones as observers and attacking weapons, camouflage is becoming a vital issue. The right camouflage suit helps you to disappear. The well-camouflaged position will remain undiscovered. There are also dummy positions and dummies that deceive the drone. When a drone is spotted, it is better to move slowly without panic, as fast movements always attract attention. If in doubt, smoke grenades deprive the drone of its view.

5. HEAT SIGNATURE MATTERS

Drones don't just see like we do, the better ones use thermal imaging. This makes it much more difficult to deal with. Essentially, it helps to have an additional surface between you and the drone: it can't look through glass because it reflects heat. A tarpaulin can protect you, or a blanket. Ultimately, the drone is just as dangerous as a sniper who works with thermal optics.

6. OBSTACLES ARE PROTECTION

It is not just the observability that needs to be considered, but also the vulnerability. FPV drones, in particular, will find their way through any opening that is large enough. Concealment of positions, closing entrances with heavy blankets, or securing them with wire mesh is necessary. And keeping an attentive look upwards prevents an enemy from suddenly hovering there and dropping something.

7. ALERTNESS IS THE FIRST DEFENSE

Alertness and attention to detail are and remain the most important self-protection measures on the battlefield. This is especially true for the threat of drones. The more conscious you are of this, the safer you are from surprises. Everyone is watching. It can also be worthwhile to assign an aerial observer. You can also detect large drones with a thermal imaging device.

8. CLOSE ENCOUNTERS OF THE DRONE KIND

Protection against micro-drones is still a tough issue. Defending against FPV drones is almost impossible because they are only visible at the last moment. A good shotgun or specially developed sighting systems for rifles and machine guns can provide a practical solution to defend against drones.

9. JAMMING WORKS BOTH WAYS

Protection against micro-drones is still a tough issue. Defending against FPV drones is almost impossible because they are only visible at the last moment. A good shotgun or specially developed sighting systems for rifles and machine guns can provide a practical solution to defend against drones.

10. STRIKE BACK

A counterattack can be an efficient form of defense: there are methods that can locate enemy drone teams. A well-coordinated team using the right technology can deliver a reactive response within three minutes using mortars or guns to combat enemy operators on the battlefield.

FURTHER READING 6

Drone defense is a very new topic, so there are still few books on the subject. Most books deal with large drones, such as the MQ-9 "Reaper", but hardly any deal with the mass use of small civilian drones that act as reconnaissance or attack aircraft, as we see in Ukraine.

▸ THE TACTICAL DRONE

Published two years ago, "The Tactical Drone" is the first manual for the use of small civilian drones (including military ones). In addition to tactical tips, it provides all kinds of suggestions for use and equipment as well as knowledge about the history of sUAS deployment.

"The Tactical Drone: Military Use of Commercial Air Vehicles — Observe / Recce / Document / Attack".
SPARTANAT Black Book 2, by Christian Väth and Markus Reisner. DGMedia, Vienna 2023 (2nd Edition), 140 pages

▶ ATP 3-01.81 COUNTER-UNMANNED AIRCRAFT SYSTEM (C-UAS)

This openly available document examines the defense against unmanned aircraft systems, such as drones and loitering munitions, at the brigade level and below. With this document, the U.S. Army responded to the changed threat situation, which makes long-lasting, high-intensity conflicts in the alliance environment more likely, and for which armed forces must urgently prepare. The first military service regulation on the subject.

"ATP 3-01.81 Counter-Unmanned Aircraft System (C-UAS)"
Headquarters, Department of the Army, Washington 2023, 58 pages. Free download: irp.fas.org/doddir/army/atp3-01-81.pdf

---

▶ A COMPREHENSIVE APPROACH TO COUNTERING UNMANNED AIRCRAFT SYSTEMS

Over the last few decades, unmanned aircraft systems (UAS) have been used in all military sectors, from small-scale micro UAS to medium-sized tactical systems and large remotely piloted aircraft (RPA). At the same time, the civil market has seen an exponential proliferation of smaller systems for public and recreational use. The best drone defense handbook to date, published by NATO's Joint Air Power Competence Centre and available for free download, provides a compact overview.

"A Comprehensive Approach to Countering Unmanned Aircraft Systems"
Joint Air Power Competence Center (JAPCC) Kalkar 2021, 644 pages. Free download: tinyurl.com/2c33ed2p

---

▸ COUNTERMEASURES FOR AERIAL DRONES

This comprehensive work describes the evolution of drones, drone threats, drone defense systems and drone defense strategies, with a focus on the practical aspects of defense systems and technologies. Covering theory, technical and operational practice with insights into industry and policing, it examines the entire threat landscape and defense technologies and systems.

"Countermeasures for Aerial Drones"
by Garik Markarian and Andrew Staniforth. Artech House, Boston and London 2020, 270 pages

---

▸ ROBOTIC WARS [GERMAN LANGUAGE]

The technological developments of recent years have led to the creation of a large number of different military "unmanned air, ground, and maritime systems". Their capabilities are leading to a transformation of modern warfare. They are the "weapons of choice" for modern militaries in the fight against asymmetric warfare and terrorism. Colonel Markus Reisner of the Austrian Armed Forces—known for his analyses of Ukraine and co-author of this book—reports on the role of drones and robots on the battlefield of tomorrow.

"Robotic Wars. Legitimatorische Grundlagen und Grenzen des Einsatzes von Military Unmanned Systems in modernen Konfliktszenarien"
by Markus Reisner. Miles Verlag, Hamburg 2018, 392 pages

▶ WIRED FOR WAR

P. W. Singer explores the greatest military revolution since the atom bomb: the beginning of robot-assisted warfare. His thesis: We are on the threshold of a massive change in military technology that threatens to make the material of "I, Robot" and "Terminator" a reality. In a mixture of historical evidence and interviews, Singer shows how technology is not only changing the way wars are fought, but also the politics, economics, laws and ethics that surround war itself. For those who like it detailed: great perspectives from an excellent author.

"Wired for War. The Robotics Revolution and Conflict in the 21st Century"
by Peter Singer. Penguin Books, New York 2011, 499 pages

▶ DRONES AND TERRORISM

Nicholas Grossman shows how we are entering the age of drone terrorism—Hezbollah squads are already using them in the Middle East. Grossman analyzes how the U.S., Israel, and other sophisticated militaries are using aerial drones and ground robots to combat non-state actors (e.g. ISIS, al-Qaeda, Iraqi and Afghan insurgents, Hezbollah, Hamas, etc.), and how such squads and individual terrorists are using less advanced, commercially available drones to attack powerful state adversaries.

"Drones and Terrorism. Asymmetric Warfare and the Threat to Global Security"
by Nicholas Grossman. I.B. Tauris, New York and London 2018, 240 pages

LEARN MORE

In addition to books, it is mainly current magazine articles that convey knowledge about the use of drones. Often it is specialist magazines that reflect on the topic. We have collected important contributions (URLs have been abbreviated):

"Make Every Marine a Drone Killer", U.S. Naval Institute: **tinyurl.com/24qpcrfr**

"Counter-Small Unmanned Aircraft Systems Strategy", U.S. Department of Defense: **tinyurl.com/24v6wxt2**

"Here's the counter-drone platforms now deployed in Ukraine", C4ISRNET: **tinyurl.com/278t3988**

"Counter-Unmanned Aircraft Systems (C-UAS)", U.S. Department of Homeland Security: **tinyurl.com/236nhj4x**

"Drohnenabwehr", Fraunhofer Gesellschaft: **tinyurl.com/24fnnlpp**

"10 Types of Counter-drone Technology to Detect and Stop Drones Today", Robin Radar: **tinyurl.com/2z9s39kf**

"Countering tactical kamikaze drones—ideas urgently needed", Wavell Room: **tinyurl.com/ytv6fvya**

ABOUT THE AUTHORS

GUSTAV FREIMANN

was a pioneer sergeant in the German Bundeswehr and, after a civilian career, joined the International Legion in Ukraine in February 2022, initially as an instructor, then as a member of the special forces, and currently as an explosive ordnance disposal expert. He shares with us his practical experience with drone defense on the front lines in Ukraine.

KRISTÓF NAGY

is a former infantryman in the German army. In addition to his work in the defense industry, he has been working as a freelance specialist author for over a decade. He is interested in kinetic effectors of all kinds and military history topics. Since 2012 he has also been working on unmanned aerial vehicles and ground systems and their defense.

MARKUS REISNER

Colonel in the General Staff Service, officer in the Austrian Federal Army, doctorate in history and PhD in law from the University of Vienna; repeated foreign assignments in Bosnia and Herzegovina, Kosovo, Afghanistan, Iraq, Chad, Central Africa and Mali. Research focus: use and future of unmanned reconnaissance and weapon systems, historical and current military issues. Currently head of the Institute for Officer Training at the Theresian Military Academy.

CHRISTIAN VÄTH

founded "Light Infantry International" in 2023. The former infantry officer was influenced above all by his stints with the Royal Marines (UK), the Telemark Battalion (NOR) and in Kabul. He is the author of Black Book 2, "The Tactical Drone". *www.lightinfantry.de*

English translation:

LAWRENCE HOLSWORTH

is a former infantry squad leader in the U.S. Army 82nd Airborne Division. In addition to his career in the soldier systems and tactical gear industry, he is also a writer and blogger. He holds a bachelor's degree in International Affairs with a focus on German Language and European Studies.

Printed in the USA
CPSIA information can be obtained
at www.ICGtesting.com
CBHW061621131024
15804CB00025B/359